D0895416

Graduate Texts in Mathematics

62

M.I. Kargapolov
Ju.I. Merzljakov

Fundamentals of
the Theory of Groups

Translated from
the Second Russian Edition by
Robert G. Burns

Springer-Verlag
New York Heidelberg Berlin

M.I. Kargapolov
formerly of
Institute of Mathematics
Novosibirsk, 90
U.S.S.R.

Ju.I. Merzljakov
Institute of Mathematics
Novosibirsk, 90
U.S.S.R.

Robert G. Burns
Translator
Department of Mathematics
York University
Downsview, Ontario M3J 1P3
Canada

Editorial Board

AMS Classification: 20-01

Kargapolov, Mikhail Ivanovich.
     Fundamentals of the theory of groups.
     (Graduate texts in mathematics; 62)
     Translation of Osnovy teorii grupp.
     Bibliography: p.
     Includes indexes.
     1. Groups, Theory of. I. Merzliakov, IUrii
Ivanovich, joint author. II. Title. III. Series.
QA171.K3713   512'.22   79-10239

The original Russian edition *Osnovy Teorii Grupp* was
published in 1977 by Nauka, Moscow.

ISBN 0-387-90396-8 Springer-Verlag   New York

ISBN 0-540-90396-8 Springer-Verlag   Berlin   Heidelberg

Preface to the Second Edition

The present edition differs from the first in several places. In particular our treatment of polycyclic and locally polycyclic groups—the most natural generalizations of the classical concept of a finite soluble group—has been expanded.

We thank Ju. M. Gorčakov, V. A. Čurkin and V. P. Šunkov for many useful remarks.

The Authors
Novosibirsk,
Akademgorodok,
January 14, 1976.

Preface to the First Edition

This book consists of notes from lectures given by the authors at Novosibirsk University from 1968 to 1970. Our intention was to set forth just the fundamentals of group theory, avoiding excessive detail and skirting the quagmire of generalizations (however a few generalizations are nonetheless considered—see the last sections of Chapters 6 and 7). We hope that the student desiring to work in the theory of groups, having become acquainted with its fundamentals from these notes, will quickly be able to proceed to the specialist literature on his chosen topic.

We have striven not to cross the boundary between abstract and scholastic group theory, elucidating difficult concepts by means of simple examples wherever possible. Four types of examples accompany the theory: numbers under addition, numbers under multiplication, permutations, and matrices. For understanding the basic text, knowledge gained from a general course in algebra will suffice; more special facts are used at times in the examples. The examples and exercises are in part used in the basic text, so that a reading of their statements should not be omitted, nor their solution postponed for too long. Solutions are included with some of these exercises. We were guided in our nomenclature by the principle of a reasonable minimum of basic terms, which required small departures from the prevailing terminology—these are noted at the appropriate places in the text.

The bibliography contains mostly group-theoretical surveys and monographs. A few references to journal articles are given immediately in the text and in general are rather random (a complete bibliography of group theory would have several thousand entries).

In a few places unsolved problems are mentioned. A rather complete collection of such problems, reflecting the interests of a wide circle of specialists in group theory, can be found in the latest edition of the "Kourovka Notebook".

The first version of this book was published in Issues 3 and 4 of the duplicated series "Library of the Department of Algebra and Mathematical Logic of NGU". We offer heartfelt thanks to all who communicated their observations to us, in particular to Ju. E. Vapne, V. D. Marzurov, V. N. Remeslennikov, N. S. Romanovskiĭ, A. I. Starostin, S. N. Černikov, and V. A. Čurkin.

The Authors
Novosibirsk,
Akademgorodok,
February 3, 1971

Translator's Remarks

1.   In his paper [Infinite groups with cyclic subgroups. Doklady Akad. Nauk SSSR **245**, No. 4 (1979)] A. Ju. Ol'šanskiĭ has announced a construction of an infinite 2-generator group all of whose proper subgroups are cyclic of prime order (where the set of primes occurring as orders is infinite). This solves at one blow Šmidt's problem (p. 14), the maximal problem (p. 137), and the minimal problem (p. 139). (Ol'šanskiĭ has also constructed a nonabelian 2-generator group, all of whose proper subgroups are infinite cyclic.) The details will appear soon in Izestija Akad. Nauk SSSR.

2.   It may be useful to explain the various notations for functions (or maps) used in the text. Let S denote a set, s an element of it, and ϕ a map with domain S. The "exponential" notation S^ϕ, s^ϕ for the images of S, s, is used only when S is being considered as a multiplicatively written group, and ϕ is a homomorphism. If S is an additive group, the notation $S\phi$, $s\phi$ is used instead. If ϕ is not primarily a group homomorphism, then the notations $S\phi$, $s\phi$; $\phi(S)$, $\phi(s)$, are used variously.

In the Russian editions the authors had introduced improvements to the conventional terminology. Unfortunately, this went almost unnoticed by the translator, so that the English terminology used is more standard. There may however be some point in mentioning a few of the authors' original terms: thus, for example, they used "period" for "exponent", "automorphically invariant" for "characteristic", and a single word (meaning "step") for "class" (of nilpotency) and "length" (of solubility).

3.   I take the opportunity of thanking Janis Leach for her excellent typing, and Maxine Burns and Abe Shenitzer for their kind advice and encouragement. Support from the National Research Council of Canada is also gratefully acknowledged.

R. G. Burns,
York University,
Toronto,
August 10, 1979

Contents

Introduction

Why does a square seem to us a symmetrical figure, a circle even more symmetrical, but the numeral "4" completely asymmetrical? To answer this question, let us consider the motions leaving each of these figures in the same place as before. It is easy to see that for the square there are eight such motions, for the circle infinitely many, but for the numeral "4" only one, the identity, which leaves each point of the numeral fixed. The set G of different motions leaving a given figure occupying the same space as before serves as a measure of its degree of symmetry: the more numerous the elements of G, i.e. the motions, the more symmetrical the figure. We define on the set G a rule of composition of its elements (or "operation") as follows: if x, y are two motions from G, then the result of composing them (called their "product" and written xy) is defined to be the motion equivalent to the successive application first of the motion x and then of the motion y. For example if x, y are the reflections of a square in its diagonals, then xy is the rotation about its centre through 180°. This composition of the elements of G clearly has the following properties: (1) $(xy)z = x(yz)$ for all elements x, y, z from G; (2) there exists in G an element e such that $xe = ex = x$ for all x from G; (3) for each x from G there exists an element x^{-1} in G such that $xx^{-1} = x^{-1}x = e$. In fact it is obvious that for e we may take the identical (or "trivial") motion, and for x^{-1} the motion opposite to x, i.e. returning each point of the figure from its new position to its old one.

Let us now leave aside our examples and consider an arbitrary set G on which an operation is given; i.e. for each two elements x, y in G there is defined an element xy again in G. If this operation satisfies conditions (1), (2), (3), then the set G with the given operation is called a *group*. Groups are basic among algebraic systems, and the theory of groups is basic among the various subdisciplines of modern algebra.

It required the work of several generations of mathematicians, spanning in all about a hundred years, before the concept of a group had crystallized out with its present clarity. In the context of the theory of algebraic equations the course of development of the group concept can be traced from Lagrange, who, in essence, applied groups of permutations to the solution of algebraic equations by radicals (1771), through the work of Ruffini (1799) and Abel (1824), to Évariste Galois, in whose work (1830) the group concept is used quite explicitly (it was he who first used the name). Independently, and for other reasons, the group concept made its appearance in geometry when in the mid-19th century the single geometry of antiquity gave way to a multitude of geometries, and the question arose of establishing the relationships between these new geometries and of classifying them. The answer was provided by the *Erlanger Programm* of Klein (1872), which proposed the idea of a group of transformations as the basis for a classification of geometries. A third source of the group concept was number theory; here among the instigators we mention only Euler, with his remainders (or "residues") after division of powers (1761), and Gauss with his composition of binary quadratic forms (1801).

The realization at the end of the 19th century that the group-theoretical ideas existing up till then independently in various areas of mathematics were essentially the same, led to the formation of the modern abstract concept of a group (by Lie, von Dyck, and others), and so to one of the earliest instances of an abstract algebraic system. This abstract group concept served in many ways as a model for the reworking, at the turn of the century, of other areas of algebra, and of mathematics generally: for these areas the process was then not so tortuous or difficult. The study of groups without the assumption of finiteness, and entirely without assumptions as to the nature of their elements, was formally inaugurated as an independent branch of mathematics with the appearance in 1916 of O. Ju. Šmidt's book "The Abstract Theory of Groups".

At the present time, group theory is one of the most highly developed branches of algebra, with numerous applications both within mathematics and beyond its boundaries: for instance to topology, function theory, crystallography, quantum mechanics, among other areas of mathematics and the natural sciences. In addition the theory has an independent life of its own, whose ultimate goal is the description of all possible group operations.

We shall now give some examples of applications of groups in algebra, in mathematics generally, and in the natural sciences.

1. *Galois groups.* Classical Galois theory consists in the application of group theory to the study of fields in the following way. Let K be a finite, separable and normal extension of a field k. The automorphisms of the field K leaving fixed the elements of the subfield k, form a group under composition of functions. This group is called the *Galois group (G say) of the extension K/k.* The fundamental theorem of Galois theory asserts that if we

associate with each subgroup $H \leq G$ its fixed subfield

$$K^H = \{x \mid x \in K, xh = x \quad \text{for all } h \in H\},$$

we obtain an anti-isomorphism of the lattice of subgroups of G onto the lattice of subfields intermediate between k and K. The field extension K^H/k will be normal if and only if the subgroup H is normal in G, and then the restriction to K^H of the automorphisms in G will yield a homomorphism, with kernel H, of the group G onto the Galois group of the extension K^H/k.

The application to the question of the solubility of equations by radicals can then be described as follows. Let f be a polynomial in x over the field k, and K the splitting field of f. Let G be the Galois group of the extension K/k. This group is also called the *Galois group of the polynomial f* over the field k. (Its elements are represented in the natural way as permutations of the roots of the equation $f(x) = 0$.) It turns out that the equation $f(x) = 0$ is soluble by radicals if and only if the Galois group of the polynomial f is soluble.

Analogous to Galois theory is the Picard-Vessiot theory in which groups are used to study extensions of differential rings and where, in particular, the question of the solubility by quadratures of differential equations is resolved. The role which in Galois theory is played by permutation groups, is in the Picard–Vessiot theory assumed by algebraic groups of matrices.

In these examples groups arise as groups of automorphisms of mathematical structures. Not only is this one of the most important ways in which they occur, but also, generally speaking, this guise is peculiar to groups and secures for them a special position in algebra. The reason for this is that one may always, in the words of Galois, "group" the automorphisms of any structure, while it is only in special cases that a ring structure or some other useful structure can be defined conveniently on the set of automorphisms.

2. *Homology groups.* The central idea of homology theory involves the application of the theory of (abelian) groups to the study of the category of topological spaces. With each space X is associated a sequence of abelian groups $H_0(X), H_1(X), \ldots$, and with each continuous map $f: X \to Y$, a sequence of homomorphisms $f_n: H_n(X) \to H_n(Y), n = 0, 1, 2, \ldots$. The study of the homology groups $H_n(X)$ and their homomorphisms by the methods of group theory often allows the solution of problems originally topological in nature. A typical example is the extension problem: Can a map $g: A \to Y$, defined on a subspace A of the space X be extended to all of X; i.e. can g be expressed as the composite of the inclusion map $h: A \to X$, and some continuous map $\hat{g}: X \to Y$? If the answer is yes, then by homology theory we must have $g_n = \hat{g}_n h_n$, i.e. each homomorphism $g_n: H_n(A) \to H_n(Y)$, can be factored through $H_n(X)$, with the factor h_n given. If this algebraic problem has a negative solution then, according to the theory, so does the original topological problem.

With this method important positive results can be obtained. By way of illustration we sketch a proof of Brouwer's fixed-point theorem: Every continuous map f of the n-dimensional ball E^n to itself has a fixed point. Suppose, on the contrary, that $f(x) \neq x$ for all $x \in E^n$. Suppose the half-line beginning at $f(x)$ and passing through the point x meets the sphere S^{n-1} (the boundary of E^n) at the point $g(x)$. Obviously g is continuous, and restricts to the identity map on S^{n-1}. Therefore the identity map on S^{n-1} can be extended to a continuous map $E^n \to S^{n-1}$. For $n = 1$ this gives a contradiction at once. If for $n \geq 2$ we compute the homology groups with coefficients from the group $\mathbf{Z}$ of integers, we find that $H_{n-1}(E^n) = 0$, $H_{n-1}(S^{n-1}) = \mathbf{Z}$, $h_{n-1} = 0$, $g_{n-1} = 1$, whence it is clear that the answer to the corresponding algebraic problem is in the negative, yielding a second, and final, contradiction.

This example from homology theory illustrates a typical mode of application of algebra (in particular group theory) to the study of non-algebraic objects: properties of the latter are elicited with the aid of algebraic systems (in particular groups) which mirror some of their structure. Such is the basic technique of algebraic topology. In the last few decades analogous techniques have been evolved, and used successfully, for studying algebraic systems themselves (for example in the theory of group extensions).

3. *Symmetry groups.* As mentioned above, the group concept allows us to give a precise meaning to the formerly slightly vague idea of the symmetry of a geometrical figure. Using this sort of approach E. S. Fedorov (1890) solved the problem, fundamental for crystallography, of classifying the regular arrangements, or lattices, of points in the Euclidean plane and in space. There turned out to be altogether just 17 planar Fedorov groups, which he discovered immediately, and 230 spatial Fedorov groups, the exhaustive classification of which relied in an essential way on group theory. This represented the first direct application of group theory to the natural sciences.

Group theory plays an analogous role in physics. Thus in quantum mechanics the state of a physical system is represented by a point of an infinite-dimensional vector space. If the system undergoes a change of state then its representing point is subjected to a certain linear transformation. Here, in addition to considerations of symmetry, the theory of representations of groups by linear transformations is important.

These examples illustrate the classifying role played by groups wherever symmetry is involved. In questions of symmetry one is dealing essentially with automorphisms of structures (not necessarily mathematical), so that in such questions group theory is irreplaceable. In mathematics itself this classifying function is of great utility: of this Klein's *Erlanger Programm* is sufficient testimony.

To summarize: the group concept, fundamental in modern mathematics, is a highly versatile tool for mathematics itself: it is used as an important constituent of many algebraic systems (e.g. rings, fields), as a sensitive register of the properties of various topological objects, as a proving-ground for the theory of algorithmic decidability, and in many other ways. It provides, in addition, a sensitive instrument for investigating symmetry, one of the most pervasive and elemental phenomena of the real world.

We conclude by listing some of the more important classes of groups.

The oldest branch of group theory, which is nonetheless developing as intensively now as it ever did in the past, is the theory of finite groups. In this theory the predominant activity at present is the search for finite *simple* groups: these embrace many of the classical groups of matrices over fields, several series of groups of automorphisms of Lie algebras, and certain isolated, "sporadic" groups. At the opposite end of the spectrum we have the finite *soluble* groups, where interest is usually concentrated on specific systems of subgroups (Hall, Carter subgroups, etc.), determining in large measure the structure of the group itself. Finite groups often arise as groups of permutations, or as matrices over finite fields; a large, and to some extent independent, segment of finite group theory occupies itself with the study of representations of groups by permutations and matrices.

In the theory of infinite groups the technique of broadest application consists in the imposition of one or another "finiteness condition". Among the classes resulting from the myriad such conditions the following come in for most attention: periodic groups, locally finite groups, groups with the maximal condition on subgroups, groups with the minimal condition on subgroups, finitely generated groups, groups of finite rank, and residually finite groups.

In abelian group theory the leading roles are played by the classes of: divisible abelian groups, torsion-free abelian groups, and by periodic abelian groups and their pure and primary subgroups. The study of general abelian groups reduces in large measure to applications of the theories of these particular classes and the theory of extensions of abelian groups, the methodology of which is largely homological in nature.

The classes of nilpotent and soluble groups, larger than that of abelian groups, can also boast of highly developed theories. Of the teeming generalizations of nilpotence and solubility we mention only: local nilpotence, the normalizer condition, the Engel condition, and the multitude of classes of groups defined by the possession of a subnormal system of one kind or another.

Several important classes of groups are obtained by imposing additional structures linked in some way to the group operation. Under this head fall, for instance, topological groups, Lie groups, linear groups and orderable groups.

Of the remaining classes we make mention of only: groups free in some variety, divisible (non-abelian) groups, groups having some property residually, automorphism groups of various mathematical structures, groups determined by conditions on their generators and defining relations, and groups with prescribed subgroup-lattices.

Definition and Most Important Subsets of a Group 1

§1. Definition of a Group

1.1. Axioms. Isomorphism

Every mathematical theory reduces ultimately to the study of two kinds of objects: sets and functions on sets. If the arguments of a function f run through a set M, in which the function also takes its values, then f is called an *algebraic operation* on M. That study which concerns itself with algebraic operations is called *algebra*. Viewed this way, algebra is concerned only with how one or another algebraic operation acts, and not at all with the set on which it is defined. The concept of isomorphism allows us to shift attention from the second of these concerns and concentrate on the first. Suppose two sets are given, together with one or more operations on each, and that there is a one-to-one correspondence between the sets themselves, and also between the sets of operations on them, such that corresponding operations are functions of the same number of variables and take corresponding values when the variables are assigned corresponding values. The sets with their operations are then said to be *isomorphic*. Isomorphic objects have identical structures as far as their operations are concerned, so that in algebra they are either not distinguished or else are regarded as exact copies of each other—much as we regard copies of a novel as being the same, even though printed with different types and on different paper, if we are interested only in the content. It makes sense to regard each class of isomorphic objects as exactly determining a certain type of algebraic operation. This reduces the problem of algebra—the study of algebraic operations—to the more concrete problem of the study of sets with operations with accuracy only up to isomorphism.

Certain kinds of algebraic operation are met with so frequently in mathematics that they have become the objects of study of independent theories. One such is the operation defining the group concept—the object of study of the theory of groups. A group is a set with one binary (i.e. two-variable) operation, satisfying certain axioms. The value of a binary operation f on a pair of elements x, y is more conveniently written, not as $f(x, y)$ as for other functions, but as xfy—this notation economises on symbols and accords well with the usual notation for numerical operations: after all we write $2+3=5$, and not $+(2, 3)=5$. In a group the binary operation is generally called multiplication and denoted by a dot (which is almost always omitted); more rarely, $+, \circ, *$, and other symbols are used. The dot notation is sometimes also referred to as the multiplicative notation, while that employing the plus sign is called the additive notation.

1.1.1. Definition. A set G with a binary operation $\cdot$ is called a *group*, if:
1. the operation is *associative*; i.e. $(ab)c = a(bc)$ for all a, b, c in G;
2. the operation guarantees an identity element; i.e. in G there is an element e—called the *identity element*—such that $ae = ea = a$ for all a in G;
3. the operation guarantees inverse elements; i.e. for each a in G there is in G an element x—called the *inverse* of a—such that $ax = xa = e$.

1.1.2. Definition. A set G with binary operation $\cdot$ is called a *group*, if
1. the operation is associative;
2. the operation guarantees left and right quotients; i.e. for each pair of elements a, b in G there are G elements x, y—called respectively *left* and *right quotients* of b by a—such that $ax = b$, $ya = b$.

1.1.3. Exercise. Definitions 1.1.1 and 1.1.2 are equivalent. The identity element of any group G is unique. Each element a in G has a unique inverse (denoted by a^{-1}). For each pair of elements a, b in G both quotients of b by a are unique. (We write $a\backslash b$ for the left quotient, and b/a for the right quotient.)

In accordance with the usual group-theoretic terminology we call a one-to-one product preserving mapping ϕ from one group onto another an *isomorphism*. In other words a map from a group G to a group G^* (in symbols $\phi: G \to G^*$) is an isomorphism, if, firstly, distinct elements have distinct images; i.e. writing a^ϕ for the image of a under the map ϕ,

$$a^\phi \neq b^\phi \text{ whenever } a \neq b, \qquad a, b \in G,$$

secondly, every element of G^* has the form g^ϕ for some $g \in G$, and, finally, the image of a product is the product of the images;

$$(ab)^\phi = a^\phi b^\phi.$$

The two groups are then said to be *isomorphic* (in symbols $G \approx G^*$).

For example, the set G of positive real numbers is a group under the usual multiplication of numbers; the set G^* of all real numbers is a group under the usual addition of numbers; and the map $\phi: G \to G^*$, defined by the formula $a^\phi = \log a$, is an isomorphism between G and G^*. When we use a logarithmic slide-rule we are simply reaping the benefits of this isomorphism. The concern of group theory is to study group operations, or, what amounts to the same thing, groups up to isomorphism. The theory of groups would be complete once a catalogue of all possible groups up to isomorphism were compiled. Happily for group theory, but unhappily for its applications, the compilation of such a catalogue is in practice impossible.

1.2. Examples

Thanks to the associative law for groups the element $(ab)c = a(bc)$ may be written simply as abc; for the same reason the product $a_1 a_2 \cdots a_n$ of n elements—without bracketing but in the given order—is uniquely defined. The product of n elements all equal to a is called the nth power of the element a, and is denoted by a^n. For zero and negative integers n we define $a^0 = e$, $a^n = (a^{-n})^{-1}$ or $a^n = (a^{-1})^{-n}$, which as it is easy to see, are equivalent.

1.2.1. Exercise. If a is any element of a group and m, n are integers, then $a^m a^n = a^{m+n}$, $(a^m)^n = a^{mn}$.

It may happen that $a^n = e$ for some $n > 0$, in which case, if $a \neq e$, the smallest n with this property is called the *order* or *period* of the element a and is denoted by $|a|$. If $a^n \neq e$ for every $n > 0$, the element a is ascribed infinite order and we write $|a| = \infty$.

1.2.2. Exercise. If $a^n = e$ then $|a|$ divides n.

1.2.3. Exercise. If the elements a, b commute, i.e. $ab = ba$, and their orders are relatively prime, then $|ab| = |a| \cdot |b|$.

1.2.4. Exercise. Suppose elements a, b commute and have orders m, n. Then the group contains an element—not always the product ab—whose order is the lowest common multiple of m and n.

We say that a group G is *torsion-free* if every nonidentity element of G has infinite order. If on the other hand every element of G has finite order then we say that G is *periodic*. If the orders of all the elements of a periodic group are bounded, then the lowest common multiple of their orders is called the *exponent* of the group. Let p be a prime. If the orders of all the elements of a periodic group are powers of p, then we call the group a *p-group*. The cardinal $|G|$ of the group G is called the *order* of G. If this cardinal is finite then we say that the group is *finite*; and in the contrary case *infinite*. If the operation in the group G is commutative, i.e. $ab = ba$ for all a, b in G, then it is said to be *commutative* or *abelian*, in honor of N. H. Abel. Often commutative operations are written additively in which case the terminology and notation are changed in accordance with the following glossary:

·	+
multiplication	addition
product	sum
identity element	zero
inverse	negative
power	multiple
quotient	difference
e or 1	0
a^{-1}	$-a$
a^n	na
$a \backslash b = a^{-1}b, \; b/a = ba^{-1}$	$b - a$

Groups are ubiquitous: Galois theory and the theory of differential equations, algebraic topology and the classification of the elementary particles of nuclear physics, crystallography and the theory of relativity, knot theory and various other branches of topology and the theory of functions—these form an incomplete list of those areas of science where the group concept makes its appearance, and moreover not just to pose decoratively but to do business. There are hefty tomes devoted to the applications of group theory in various branches of science and to these we refer the interested reader. Here we give only a few examples taken from the material of a course in general algebra—the kind of material to which we limit ourselves in this book. (The symbols in heavy type will throughout the book be used consistently for the appropriate "classical" objects. They are all standard notation—even down to their nuances of meaning—and are widely used in the literature; there is an index of these symbols at the end of the book.)

1.2.5. EXAMPLES. (I). The set of all elements of an arbitrary ring K, under the additive operation of the ring, is an abelian group. It is called the *additive group of the ring K*. We shall use the same letter K to denote it, adhering to the general convention that capital letters denote sets and that it should be clear from the context which operations on the sets are under consideration. In particular the additive groups of the field $\mathbf{C}$ of complex numbers, the field $\mathbf{R}$ of real numbers, the field $\mathbf{Q}$ of rational numbers and the ring $\mathbf{Z}$ of integers, are all torsion-free abelian groups. More generally, the additive group of a field of characteristic zero is torsion-free, while the additive group of a field of characteristic p has exponent p. The additive groups of the finite field

GF(*q*) of *q* elements and the ring $\mathbf{Z}_n$ of residues modulo *n* are finite abelian groups; moreover,

$$|\mathbf{GF}(q)| = q, \qquad |\mathbf{Z}_n| = n.$$

Let *p* be a prime. Denote by $\mathbf{Q}_p$ the set of all rational numbers of the form m/p^n where *m*, *n* are integers. Under the usual addition of numbers $\mathbf{Q}_p$ is a torsion-free abelian group.

(II). The set of all invertible elements of an arbitrary ring *K* with a multiplicative identity element is a group under the multiplicative operation of *K*. It is called the *multiplicative group of the ring K*, and is denoted by K^*. (The set K^* clearly does not contain the zero and is thus different from *K*.) If the ring *K* is commutative then so is the group K^*. In particular the multiplicative groups $\mathbf{C}^*$, $\mathbf{R}^*$, $\mathbf{Q}^*$, $\mathbf{Z}^*$, $\mathbf{GF}(q)^*$, $\mathbf{Z}_n^*$ are all abelian. The set $\mathbf{C}_n$ of complex numbers satisfying the equation $x^n = 1$, with the usual multiplication of numbers, is an abelian group. Let *p* be a prime. The set $\mathbf{C}_{p^\infty}$ of all roots of the equation $x^{p^n} = 1$, $n = 1, 2, \ldots$, in the field of complex numbers, is, under the usual multiplication, an infinite abelian *p*-group. It is called the *quasicyclic group of type* p^∞. The groups $\mathbf{Z}^*$, $\mathbf{GF}(q)^*$, $\mathbf{Z}_n^*$, $\mathbf{C}_n$, are finite; moreover,

$$|\mathbf{Z}^*| = 2, \qquad |\mathbf{GF}(q)^*| = q - 1, \qquad |\mathbf{Z}_n^*| = \phi(n), \qquad |\mathbf{C}_n| = n,$$

where ϕ is Euler's function, i.e. $\phi(mn) = \phi(m)\phi(n)$ for relatively prime *m*, *n*, and $\phi(p^k) = p^k - p^{k-1}$ for *p* prime.

(III). Let *M* be a set, and $\mathbf{S}(M)$ the set of one-to-one maps of *M* onto itself. If we define multiplication in $\mathbf{S}(M)$ to be composition of maps, then $\mathbf{S}(M)$ becomes a group. In particular, if $M = \{1, \ldots, n\}$, this group is just the group of all permutations of degree *n*, and is called the *symmetric group of degree n*, denoted by $\mathbf{S}_n$. The group $\mathbf{S}_n$ is finite of order *n*!. For $n > 2$ the group $\mathbf{S}_n$ is nonabelian.

(IV). The set $\mathbf{GL}_n(K)$ of all invertible matrices of degree *n* over a commutative ring *K* with an identity is a group under the usual multiplication of matrices. It is called the *general linear* or *general matrix group* of degree *n* over the ring *K*. It is clear that $\mathbf{GL}_n(K)$ is just $\mathbf{M}_n(K)^*$, where $\mathbf{M}_n(K)$ is the ring of all matrices of degree *n* over the ring *K*. For $n \geq 2$ the group $\mathbf{GL}_n(K)$ is nonabelian. Consider the following subsets of $\mathbf{GL}_n(K)$: $\mathbf{SL}_n(K)$, the subset consisting of all matrices with determinant 1; $\mathbf{D}_n(K)$, the subset of diagonal matrices; $\mathbf{T}_n(K)$, the subset of matrices with all entries below the main diagonal zero; and $\mathbf{UT}_n(K)$, the subset of matrices with all entries below the main diagonal zero, and with the entries on the main diagonal all the identity. All of these sets are also groups under matrix multiplication. They are called, respectively: the *special linear group*, the *diagonal group*, the (upper) *triangular group*, and the *unitriangular group*. In the case that *K* is the finite field $\mathbf{GF}(q)$, in place of $\mathbf{GL}_n(K)$ we usually write $\mathbf{GL}_n(q)$, and similarly for the other matrix groups.

1.2.6. Exercise. $\mathbf{Z}_n \simeq \mathbf{C}_n$.

1.2.7. Exercise. The group $\mathbf{C}^*$ is isomorphic to the group of all nonsingular matrices of the form

$$\begin{pmatrix} \alpha & \beta \\ -\beta & \alpha \end{pmatrix}$$

with real entries, under matrix multiplication.

1.2.8. Exercise. $\mathbf{Q}_p \neq \mathbf{Q}_q$ if $p \neq q$.

1.2.9. Exercise. If a permutation a in $\mathbf{S}_n$ is a product of disjoint cycles of lengths $n_1, \ldots, n_k$, then its order $|a|$ is the lowest common multiple of the numbers $n_1, \ldots, n_k$. (A subset of $\mathbf{S}_n$ consists of *disjoint* permutations if the subsets of $\{1, \ldots, n\}$ consisting of the elements moved (i.e. not fixed) by the permutations, are pairwise disjoint.)

Our reserve of examples of groups will be greatly enlarged if we can find ways of constructing new groups from given ones. Group theory has in its arsenal a variety of such constructions. One of the simplest, but at the same time most important, constructions consists in the following.

Let $G_1, \ldots, G_m$ be groups. It is easy to verify that the set $G = G_1 \times \cdots \times G_m$ of sequences $(g_1, \ldots, g_m)$, $g_i \in G_i$, with componentwise multiplication

$$(g_1, \ldots, g_m) \cdot (g'_1, \ldots, g'_m) = (g_1 g'_1, \ldots, g_m g'_m),$$

is a group. It is called the *direct* or *Cartesian product* of the groups G_i, and the G_i are called (direct) *factors* of G. It is easy to extend this concept to the situation of an arbitrary collection of factors G_α, $\alpha \in I$. Thus we denote by

$$G = \prod_{\alpha \in I} G_\alpha$$

the set of functions

$$f: I \rightarrow \bigcup_{\alpha \in I} G_\alpha$$

satisfying the condition that $f(\alpha) \in G_\alpha$ for all $\alpha \in I$. It is readily checked that the set G with multiplication defined by the rule $(fg)(\alpha) = f(\alpha)g(\alpha)$, is a group; it is also called the *Cartesian product* of the groups G_α. The value of a function f at the element α is called the *projection* of f on the factor G_α, or the *component* of f in G_α. The set

$$\text{supp } f = \{\alpha \mid \alpha \in I, f(\alpha) \neq e\}$$

is called the *support* of the function f. Clearly the set of functions with finite support in the Cartesian product of the groups G_α, is itself a group under the same multiplication. We shall call this group the *direct product* of the groups

G_α, and denote it by $\prod_{\alpha \in I}^{\times} G_\alpha$. Obviously, as has already been indicated, for a finite number of factors the direct and Cartesian products coincide.

In additive terminology instead of products we speak of sums, instead of factors, summands, and we write

$$G = G_1 \oplus \cdots \oplus G_m,$$

$$G = \sum_{\alpha \in I} G_\alpha, \qquad G = \bigoplus_{\alpha \in I} G_\alpha.$$

1.2.10. Exercise. $\mathbf{C}_m \times \mathbf{C}_n = \mathbf{C}_{mn}$ for relatively prime m, n.

1.2.11. Exercise. $\mathbf{D}_n(K) \simeq K^* \times \cdots \times K^*$ (n times).

§2. Subgroups. Normal Subgroups

2.1. Subgroups

If a subset H of a group G is closed under the group operation, i.e. together with any two of its elements a, b, contains also their product ab, then the restriction of the operation to H will be an algebraic operation on H; we say that this operation is *induced* by the operation on G. If H turns out to be a group under the induced operation, it is called a *subgroup* of the group G, and we write $H \leq G$. If $H \leq G$ and $H \neq G$, we write $H < G$. (Do not confuse these symbols with the symbols $\subseteq$, $\subset$ for set inclusion!)

2.1.1. Exercise. For a subset H of a group G to be a subgroup, it is necessary and sufficient that H be closed under multiplication and taking of inverses, i.e. that, together with every two of its elements a, b, it contain also ab and a^{-1}. These closure conditions may also be written as:

$$HH \subseteq H, \qquad H^{-1} \subseteq H,$$

where here we are using the usual definitions

$$AB = \{ab \mid a \in A, \, b \in B\}, \qquad A^{-1} = \{a^{-1} \mid a \in A\},$$

for subsets A, B of the given group.

2.1.2. Exercise. The product AB of subgroups A, B of a group G, is a group if and only if $AB = BA$.

2.1.3. Exercise. If A, B are finite subgroups of G, then

$$|AB| = \frac{|A| \cdot |B|}{|A \cap B|}.$$

In every group the set containing only the identity element, and the group itself, are obvious subgroups. Subgroups other than these are called *proper* subgroups.

Continuing our investigation of the groups of Examples (I) to (IV) of §1.2, we now indicate certain of their subgroups.

2.1.4. EXAMPLES. (I). Obviously,

$$\mathbf{Z} < \mathbf{Q}_p < \mathbf{Q} < \mathbf{R} < \mathbf{C},$$

$$\mathbf{Z} = \bigcap \mathbf{Q}_p,$$

$$\mathbf{GF}(p^m) \le \mathbf{GF}(p^n) \quad \text{if } m \mid n.$$

(Here we understand $\mathbf{GF}(p^k)$ to be the appropriate subset of the algebraic closure of $\mathbf{GF}(p)$.)

(II). Obviously,

$$\mathbf{Z}^* < \mathbf{Q}^* < \mathbf{R}^* < \mathbf{C}^*,$$

$$\mathbf{C}_p < \mathbf{C}_{p^2} < \cdots < \mathbf{C}_{p^\infty},$$

$$\mathbf{C}_{p^\infty} = \bigcup \mathbf{C}_{p^n},$$

$$\mathbf{GF}(p^m)^* \le \mathbf{GF}(p^n)^* \quad \text{if } m \mid n.$$

III. In the symmetric group $\mathbf{S}_n$ the subset $\mathbf{A}_n$ of all even permutations forms a subgroup called the *alternating group* of degree n. Clearly $|\mathbf{A}_n| = n!/2$.

IV. For $n \ge 2$,

$$\mathbf{SL}_n(K) \le \mathbf{GL}_n(K),$$

$$\mathbf{D}_n(K) < \mathbf{T}_n(K),$$

$$\mathbf{UT}_n(K) \le \mathbf{T}_n(K) < \mathbf{GL}_n(K).$$

We mention also the *orthogonal group*, i.e. the subgroup of $\mathbf{GL}_n(K)$ defined by

$$\mathbf{O}_n(K) = \{a \mid aa' = e\},$$

where the prime indicates the transpose; and when $K = \mathbf{C}$ the *unitary group*, i.e. the subgroup of $\mathbf{GL}_n(\mathbf{C})$ defined by

$$\mathbf{U}_n = \{a \mid a\bar{a}' = e\},$$

where the bar indicates that the matrix entries have been replaced by their complex conjugates. Finally,

$$\mathbf{UT}_n(K) = \mathbf{UT}_n^1(K) \ge \mathbf{UT}_n^2(K) \ge \cdots,$$

where $\mathbf{UT}_n^m(K)$ is the set of matrices in $\mathbf{UT}_n(K)$ with all entries zero in the $(m-1)$ diagonals immediately above the main diagonal.

2.2. Generating Sets

It is easily seen that the intersection of any collection of subgroups of a group is again a subgroup. If M is any subset of a group G, the intersection $\langle M \rangle$ of all subgroups containing M is called the subgroup *generated* by the set M, and M is termed a *generating set* for the subgroup $\langle M \rangle$. Allowing more rein to the terminology, we sometimes speak of the elements of M as *generators* of the subgroup $\langle M \rangle$. A group generated by some finite subset is said to be *finitely generated*.

2.2.1. Theorem. *If M is a subset of a group G, then*
$$\langle M \rangle = \{ a_1^{\varepsilon_1} \cdots a_m^{\varepsilon_m} \mid a_i \in M, \ \varepsilon_i = \pm 1, \ m = 1, 2, \ldots \}.$$

PROOF. Denote the right-hand side by H. Since the subgroup $\langle M \rangle$ contains all a_i in M, we have $\langle M \rangle \supseteq H$. On the other hand $HH \subseteq H$, $H^{-1} \subseteq H$, so that H is a subgroup containing M. Hence $H \supseteq \langle M \rangle$, and finally $H = \langle M \rangle$.

By way of illustration we give generating sets for the groups of Examples (I) to (IV) of §1.2. For ease of notation we shall write
$$\langle M \rangle = \langle \cdots \mid \cdots \rangle,$$
if M is given in the form
$$M = \{ \cdots \mid \cdots \},$$
i.e. we shall omit the braces.

2.2.2. EXAMPLES. (I). Obviously,
$$\mathbf{Z} = \langle 1 \rangle,$$
$$\mathbf{Z}_n = \langle 1 (\bmod n) \rangle,$$
$$\mathbf{Q} = \left\langle \frac{1}{n} \,\middle|\, n = 1, 2, \ldots \right\rangle.$$

(II). Obviously,
$$\mathbf{Z}^* = \langle -1 \rangle,$$
$$\mathbf{Q}^* = \langle -1, 2, 3, 5, 7, 11, \ldots \rangle,$$
$$\mathbf{GF}(q)^* = \langle \zeta_q \rangle,$$
$$\mathbf{C}_n = \langle \alpha_n \rangle,$$
$$\mathbf{C}_{p^\infty} = \langle \alpha_{p^m} \mid m = 1, 2, \ldots \rangle,$$

where ζ_q is a primitive root of the equation $x^{q-1} = 1$ over the field $\mathbf{GF}(q)$, and
$$\alpha_n = \cos \frac{2\pi}{n} + i \sin \frac{2\pi}{n}.$$

(III). From a course in general algebra it should be known that the symmetric group $\mathbf{S}_n$ is generated by its transpositions (ij). Since $(ij) = (1i)(1j)(1i)$, it follows that $\mathbf{S}_n$ is even generated by the particular transpositions $(12), (13), \ldots, (1n)$. The alternating group $\mathbf{A}_n$ is generated by the set of all 3-cycles (ijk) since an even permutation is a product of an even number of transpositions and

$$(ij)(ik) = (ijk), \qquad (ij)(kl) = (ilj)(jkl).$$

(IV). Let K be a field. Consider in $\mathbf{GL}_n(K)$ the matrix $t_{ij}(\alpha)$ defined for each $1 \le i, j \le n$, and $\alpha \in K$, by

$$t_{ij}(\alpha) = e + \alpha e_{ij},$$

where e is the identity matrix and e_{ij} is the matrix with its (i, j)th entry 1, and all other entries 0. The matrices $t_{ij}(\alpha)$ with $i \ne j$, $\alpha \ne 0$, are called *transvections*. Define also for each $0 \ne \beta \in K$ the diagonal matrix $d(\beta)$ by

$$d(\beta) = e + (\beta - 1)e_{nn}.$$

We shall now prove that each matrix in $\mathbf{GL}_n(K)$ can be expressed as a product

$$t_1 \cdots t_r d(\beta) t_{r+1} \cdots t_s,$$

where the t_i are transvections. This will imply, in particular, that

$$\mathbf{GL}_n(K) = \langle t_{ij}(\alpha), d(\beta) \mid 0 \ne \alpha, \beta \in K, i \ne j \rangle, \tag{1}$$

$$\mathbf{SL}_n(K) = \langle t_{ij}(\alpha) \mid 0 \ne \alpha \in K, i \ne j \rangle. \tag{2}$$

The promised proof is as follows. Observe that multiplication of a matrix on the right by a transvection has the same effect as changing one of its columns by adding to it a scalar multiple of some other column, while multiplication on the left by a transvection has the analogous effect on one of its rows; such changes in a matrix are called *elementary transformations*. Let $a \in \mathbf{GL}_n(K)$. By means of elementary transformations of the columns of a it can be arranged that in the resulting matrix the $(1, 2)$th entry, a_{12} say, is not zero. After a further elementary transformation, namely changing the first column by adding to it the second multiplied by $(1 - a_{11})/a_{12}$, we obtain a matrix with the identity element as its $(1, 1)$th entry, and then with the help of this entry, after further elementary transformations we obtain a matrix with zeros everywhere in its first row and column except for the $(1, 1)$th entry, where we have 1. Continuing this process, applied next (in essence) to the $(n-1) \times (n-1)$ submatrix obtained by deleting the first row and column, and so on, we arrive finally at a matrix of the form $d(\beta)$. Hence $a = t_1 \cdots t_r d(\beta) t_{r+1} \cdots t_s$, as required.

It can be proved in a similar manner that

$$\mathbf{T}_n(K) = \langle t_{ij}(\alpha), \operatorname{diag}(\beta_1, \ldots, \beta_n) \mid 0 \ne \alpha, \beta_k \in K, i < j \rangle, \tag{3}$$

$$\mathbf{UT}_n^m(K) = \langle t_{ij}(\alpha) \mid 0 \ne \alpha \in K, j - i \ge m \rangle. \tag{4}$$

The result just proved allows us to define $\mathbf{SL}_n(K)$ when K is a noncommutative ring, in which case the usual definition of determinant is inapplicable. Thus for such K simply take (2) as the definition. Further, let K be a division ring. In a way similar to the above it can be shown that every invertible matrix a over K can be written in the form $a = t_1 \cdots t_r d(\beta) t_{r+1} \cdots t_s$, $\beta \neq 0$. For the reader familiar with the concepts of commutator subgroup and quotient group, we mention that the image of the element β in the quotient of the group K^* by its commutator subgroup, is called the *determinant of the matrix a in the sense of Dieudonné*.

2.2.3. Exercise. Let $n \geq 2$, $p_1, \ldots, p_n$ be distinct primes, and $\hat{p}_i = p_1 \cdots p_{i-1} p_{i+1} \cdots p_n$. Then

$$\mathbf{Z} = \langle \hat{p}_1, \ldots, \hat{p}_n \rangle,$$

and no member of this generating set can be omitted.

2.2.4. Exercise. $\mathbf{S}_n = \langle (12), (12 \cdots n) \rangle$.

2.2.5. Exercise. $\mathbf{SL}_n(\mathbf{Z}) = \langle e + e_{ij} \mid 1 \leq i, j \leq n, i \neq j \rangle$;

$$\mathbf{SL}_n(\mathbf{Z}) = \langle e + e_{12}, e_{12} + e_{23} + \cdots + e_{n-1,n} + (-1)^{n-1} e_{n1} \rangle.$$

The concept antipodal to that of a generating set is that of the Frattini subgroup. To give the definition of this concept we first define a subgroup H of a group G to be *maximal with respect to a property* σ, if H has the property, and there is no larger subgroup with the property. If the property σ is such that all groups possess it then we simply use without qualification the term *maximal* for subgroups maximal with respect to the property σ. Of course a group may have no maximal subgroups—see the examples below. We now define the *Frattini subgroup* $\Phi(G)$ of a group G to be the intersection of all its maximal subgroups if there are any, and G itself if there are none.

An element of a group G is called a *nongenerator* if it can be omitted from every generating set for G which contains it.

2.2.6. Theorem. *The Frattini subgroup $\Phi(G)$ of a group G coincides with the set S of all nongenerators of G.*

PROOF. (i) $S \subseteq \Phi(G)$. If G has no maximal subgroups then the inclusion is obvious. Suppose H is a maximal subgroup of G and let $x \in S$. If $x \notin H$ then $\langle x, H \rangle = G$ while $\langle H \rangle \neq G$. This contradicts the fact that $x \in S$. Hence $x \in H$ and so $x \in \Phi(G)$.

(ii) $\Phi(G) \subseteq S$. Suppose on the contrary that there is an element $x \in \Phi(G)$ which together with some set M generates G, while $\langle M \rangle \neq G$. By Zorn's Lemma there exist subgroups maximal with respect to containing M and not containing x. It is clear that these subgroups are (absolutely) maximal. They thus all contain $\Phi(G)$, and therefore x, contrary to their definition. This completes the proof of the theorem.

2.2.7. EXAMPLES. (I). In the group $\mathbf{Z}$ the subgroup $\langle p \rangle$ is maximal for each prime p, so that $\Phi(\mathbf{Z}) = 0$. It is easily seen that in the group $\mathbf{Q}$ every element is a nongenerator, whence $\Phi(\mathbf{Q}) = \mathbf{Q}$.

(II). Since the group $\mathbf{C}_{p^\infty}$ is the union of its subgroups $\mathbf{C}_{p^n}$, $n = 1, 2, \ldots$, all of its elements are nongenerators. Therefore $\Phi(\mathbf{C}_{p^\infty}) = \mathbf{C}_{p^\infty}$.

(III). It can be verified that the subgroup H_i of the group $\mathbf{S}_n$, consisting of all permutations stabilizing the symbol i, is maximal in $\mathbf{S}_n$. Since the intersection $H_1 \cap \cdots \cap H_n$ is 1, we have that $\Phi(\mathbf{S}_n) = 1$. Similarly it can be shown that $\Phi(\mathbf{A}_n) = 1$.

(IV). Consider in the group $\mathbf{UT}_n(\mathbf{Z})$ the subgroup H_{ip} consisting of all matrices x with $x_{i,i+1} \in \langle p \rangle$, where $1 \le i \le n - 1$, and p is prime. It may be checked that H_{ip} is maximal in $\mathbf{UT}_n(\mathbf{Z})$. Since the intersection of all H_{ip} is contained in $\mathbf{UT}_n^2(\mathbf{Z})$, we have that $\Phi(\mathbf{UT}_n(\mathbf{Z})) \le \mathbf{UT}_n^2(\mathbf{Z})$. It may be shown that in fact the reverse inclusion holds, i.e.

$$\Phi(\mathbf{UT}_n(\mathbf{Z})) = \mathbf{UT}_n^2(\mathbf{Z}).$$

2.3. Cyclic Subgroups

A subgroup $\langle a \rangle$ generated by a single element a, is called *cyclic*. By Theorem 2.2.1 it consists of all possible powers of the generator:

$$\langle a \rangle = \{ a^n \mid n = 0, \pm 1, \pm 2, \ldots \}.$$

Example 2.2.2 (I) shows that $\mathbf{Z}$ and $\mathbf{Z}_n$ are cyclic groups. It turns out that up to isomorphism these exhaust the supply of cyclic groups.

2.3.1. Theorem. *Every infinite cyclic group is isomorphic to the group* $\mathbf{Z}$, *and every cyclic group of finite order is isomorphic to some group* $\mathbf{Z}_n$.

PROOF. Let $\langle a \rangle$ be an infinite cyclic group. Define a mapping $\phi : \mathbf{Z} \to \langle a \rangle$ by $n\phi = a^n$. It is one-to-one: if for $m > n$ we had $m\phi = n\phi$, i.e. $a^{m-n} = e$, then the group $\langle a \rangle$ would turn out to be finite. Further $(m + n)\phi = m\phi n\phi$, i.e. ϕ is an isomorphism (since it is clearly onto). If $\langle b \rangle$, $\langle c \rangle$ are cyclic groups both of finite order n, then the map defined by $b^k \to c^k$, $0 \le k \le n - 1$, is an isomorphism between $\langle b \rangle$ and $\langle c \rangle$.

2.3.2. Theorem. *Every subgroup of a cyclic group is cyclic.*

PROOF. Let $\langle a \rangle$ be a cyclic group of order n, and H a nontrivial subgroup. (It is obvious that the identity subgroup is cyclic.) Let m be the smallest positive integer such that

$$a^m \in H, \qquad 0 < m < n.$$

Obviously $\langle a^m \rangle \subseteq H$. We shall show that $\langle a^m \rangle = H$. Let a^k, $0 \le k < n$ be any element of H. Dividing m into k we get: $k = mq + r$, $0 \le r < m$. Then

$$a^r = a^k (a^m)^{-q} \in H.$$

By the choice of m it follows that $r = 0$, whence $a^k \in \langle a^m \rangle$. The cyclicity of the subgroups of the infinite cyclic group is proved similarly.

2.3.3. Exercise. A group is called *locally cyclic* if every finite subset of the group generates a cyclic subgroup. Prove that $\mathbf{Q}$ and $\mathbf{C}_{p^\infty}$ are locally cyclic. It follows that they themselves are not finitely generated.

2.4. Cosets

Given a subgroup H of a group G we can form the sets

$$gH = \{gh \mid h \in H\}, \qquad g \in G,$$

which are called *left cosets* of the subgroup H in the group G. *Right cosets* Hg are defined similarly. Each element of a coset is called a *representative* of the coset. It is easily seen that

$$aH = bH \Leftrightarrow a^{-1}b \in H, \tag{5}$$

$$Ha = Hb \Leftrightarrow ab^{-1} \in H. \tag{6}$$

These allow us to arrive at the concept of a coset by another route: Define for a given subgroup H the relation $\sim$ on G of *left congruence* by

$$a \sim b \Leftrightarrow a^{-1}b \in H,$$

and define *right congruence* analogously. It is easily verified that these relations are equivalences on G, i.e. they are reflexive ($a \sim a$), symmetric ($a \sim b \Rightarrow b \sim a$), and transitive ($a \sim b$, $b \sim c \Rightarrow a \sim c$), and therefore they yield two partitions of G. In view of (5), (6) these partitions coincide respectively with the partition into left cosets and the partition into right cosets. Hence in particular the left cosets of H are pairwise disjoint, and the same is true of the right cosets of H.

Since the correspondence $gH \leftrightarrow Hg^{-1}$ is one-to-one, the cardinal of the collection of left cosets of H is the same as that of the collection of right cosets. This cardinal is called the *index* of the subgroup H in the group G, and is denoted by $|G:H|$.

2.4.1. Exercise. $|\mathbf{Z} : \langle n \rangle| = n$.

2.4.2. Exercise. The group $\mathbf{Q}$ contains no proper subgroups of finite index.

2.4.3. Exercise. $|\mathbf{S}_n : \mathbf{A}_n| = 2$.

2.4.4. Exercise. If A, B are subgroups of a group G, then

$$|A:A\cap B|\leq|G:B|. \tag{7}$$

(Hint: The left congruence on A relative to $A\cap B$ is the restriction to A of the left congruence on G relative to B.)

Each coset gH, Hg has the same cardinal as H as is shown by the bijections $h\leftrightarrow gh$, $h\leftrightarrow hg$, $h\in H$. Thus in particular if the group G is finite then $|G|$ may be calculated by multiplying the cardinal $|H|$ of each coset by the number $|G:H|$ of cosets. In this way we obtain

2.4.5. Lagrange's Theorem. *If H is a subgroup of a finite group G, then*

$$|G|=|H|\cdot|G:H|. \tag{8}$$

This has one very important consequence: The order of a subgroup always divides the order of the group. Since $|a|=|\langle a\rangle|$, we deduce that the order of an element always divides the order of the group.

2.4.6. Exercise. Every group of prime order is cyclic.

2.4.7. Exercise. Let A, B be subgroups of a group G with $A\leq B$. The indices $|G:B|$, $|B:A|$ are both finite if and only if $|G:A|$ is finite. If the index $|G:A|$ is finite then

$$|G:A|=|G:B|\cdot|B:A|. \tag{9}$$

This generalizes Lagrange's Theorem (which corresponds to the case $A=1$).

2.4.8. Exercise. The intersection of a finite number of subgroups of finite index is again of finite index. (Hint. From (7), (9) derive the inequality

$$|G:A\cap B|\leq|G:A|\cdot|G:B|. \tag{10}$$

Alternatively, use the fact that $Ag\cap Bg=(A\cap B)g$.)

2.4.9. Exercise. The intersection of all finite index subgroups of $\mathbf{Q}_p$ is the zero subgroup. In fact the same is true for every proper subgroup of the group $\mathbf{Q}$.

2.4.10. Exercise. The groups $\mathbf{C}_{p^n}$, $n=1,2,\ldots$, comprise all proper subgroups of the quasicyclic group $\mathbf{C}_{p^\infty}$.

We see from the last exercise that all the proper subgroups of the quasicyclic groups are finite although they themselves are infinite. It is unknown whether or not there are groups other than the quasicyclic groups with this property.[†] (This is the famous problem of O. Ju. Šmidt.) It *is* known, however, that there are no such groups (apart from the quasicyclic

† There do exist such groups: see Translator's Remarks, p. ix.

groups) among the locally finite groups, which we shall meet with later [M. I. Kargapolov, Sib. matem. ž. **4**, No. 1 (1963), 232–235].

2.5. Classes of Conjugate Elements

In group theory a particularly important role is played by those subgroups whose left and right cosets coincide. Such subgroups are said to be normal. More precisely we say that a subgroup H of a group G is *normal* in G, and write $H \unlhd G$, if $Hx = xH$ for every x in G. The notation $H \lhd G$ means that $H \unlhd G$ but $H \neq G$.

It is clear that the condition $Hx = xH$ is equivalent to the condition $x^{-1}Hx = H$. We say that an element a is *conjugate* to an element b by an element x if $a = x^{-1}bx$. Often the latter is written with the conjugating element as a superscript: $x^{-1}bx = b^x$. It is easily verified that we always have

$$(ab)^x = a^x b^x, \qquad (a^x)^y = a^{xy}. \tag{11}$$

If A, B are subsets of a group we write also

$$A^B = \{a^b \mid a \in A, b \in B\}.$$

Thus, rephrasing our definition, we may say that a subgroup H of a group G is normal in G if and only if it contains all conjugates by elements of G of all of its elements, or, more briefly, if

$$H^G \subseteq G.$$

The first equation of (11) shows that, for a fixed $x \in G$, the map of the group G onto itself given by the rule $a \to a^x$, is an isomorphism. A subset M is said to be *conjugate* in G to the subset M^x. We can now rephrase the definition of normality again: a subgroup is normal if and only if it is the same as each of its conjugates. For this reason normal subgroups are also called "self-conjugate." The term "invariant subgroup" is also in use. A group is called *simple* if it has no proper normal subgroups.

2.5.1. EXAMPLES. (I), (II). Since the additive and multiplicative groups of a field are abelian, all of their subgroups are normal.

(III). Since a conjugate of an even permutation is again even, we have that $\mathbf{A}_n \lhd \mathbf{S}_n$.

(IV). Let $n \geq 2$. Since $\det(ab) = \det a \cdot \det b$, a matrix with determinant 1 goes after conjugation to a matrix with the same property; thus $\mathbf{SL}_n(K) \unlhd \mathbf{GL}_n(K)$. Since each diagonal entry in the product of two triangular matrices is the product of the corresponding entries in the factors, we have also that $\mathbf{UT}_n(K) \unlhd \mathbf{T}_n(K)$. In fact one may verify that $\mathbf{UT}_n^m(K) \unlhd \mathbf{T}_n(K)$ for all $m = 1, 2, \ldots$.

2.5.2. Exercise. If p is prime then $\mathbf{Z}_p$ is simple.

2.5.3. Exercise. If $|G:H| = 2$ then $H \lhd G$.

2.5.4. Exercise. If $A \unlhd G$, $B \unlhd G$ then $AB \unlhd G$.

2.5.5. Exercise. It may happen that $A \unlhd B$, $B \unlhd C$ but $A \ntrianglelefteq C$.

Let G be any group. We can define a relation $\sim$ on G by setting $a \sim b$ whenever a and b are conjugate in G. It is easily verified that this relation is an equivalence (see the preceding subsection), so that G is partitioned into pairwise nonintersecting *classes of conjugate elements*, or *conjugacy classes* a^G. In particular normal subgroups are thus just those subgroups which are unions of one or more complete classes of conjugate elements.

In contrast to cosets, conjugacy classes may be of various cardinals. A decisive role in computing their sizes is played by the concept of the normalizer. Let M be a subset and H a subgroup of a group G. The *normalizer* of the subset M in the subgroup H is the set

$$N_H(M) = \{h \mid h \in H, M^h = M\},$$

which, as is readily checked, is a subgroup of H. If it is not indicated in which subgroup H the normalizer is taken, then it is to be understood that it is taken in the whole group G. Obviously a subgroup is normal in a group if and only if its normalizer is the whole group.

2.5.6. Theorem. *If M is a subset and H a subgroup of a group G then the cardinal of the class of subsets conjugate with M by elements of H is equal to the index $|H:N_H(M)|$. In particular*

$$|a^G| = |G:N_G(a)|.$$

PROOF. Map the collection of sets M^x, $x \in H$, into the collection of right cosets of the subgroup $N = N_H(M)$ in H, according to the following rule:

$$M^x \to Nx, \qquad x \in H.$$

This map is single-valued since from $M^x = M^y$ follows $Nx = Ny$. It takes distinct elements to distinct elements since from $Nx = Ny$ follows $M^x = M^y$. Finally, the map is *onto* since each Nx has preimage M^x.

2.5.7. EXAMPLES. (I), (II). Since the additive and multiplicative groups of a field are abelian, every one of their conjugacy classes consists of a single element.

(III). Let M be a set. Two elements of $S(M)$ are conjugate in $S(M)$ precisely if their decompositions into disjoint cycles contain the same number of cycles of each length, including those of length 1. (Here number and length of cycles are to be understood in the sense of cardinals.) To be more explicit, if

$$a = (\alpha_1\alpha_2 \cdots)(\beta_1\beta_2 \cdots)(\cdots)\cdots,$$
$$a' = (\alpha_1'\alpha_2' \cdots)(\beta_1'\beta_2' \cdots)(\cdots)\cdots,$$

where cycles of the same length are written one directly under the other, then it is immediately verifiable that $a' = a^x$ where

$$x = \begin{pmatrix} \alpha_1\alpha_2 \cdots \beta_1\beta_2 \cdots \\ \alpha_1'\alpha_2' \cdots \beta_1'\beta_2' \cdots \end{pmatrix}.$$

In particular two permutations in $\mathbf{S}_n$ are conjugate in $\mathbf{S}_n$ if and only if they have the same disjoint cycle structure: for example the permutation $(12)(3456)$ is conjugate in $\mathbf{S}_6$ with the permutation $(15)(2436)$, but not with the permutation $(12)(345)(6)$. In $\mathbf{A}_n$ the elements with the same cycle structure fall into either a single conjugacy class or two conjugacy classes of the same cardinal—this is easily perceived using Theorem 2.5.6 and the equation $|\mathbf{S}_n : \mathbf{A}_n| = 2$.

(IV). As the reader will recall, the question of conjugacy of matrices occupies an important place in a course in general algebra. For algebraically closed fields K the question of when elements of $\mathbf{GL}_n(K)$ are conjugate is completely answered by the theorem of Jordan: two matrices in $\mathbf{GL}_n(K)$ are conjugate in that group if and only if they have the same Jordan form.

2.5.8. Exercise. The orders of conjugate elements (or, more generally, of conjugate subsets) are the same.

2.5.9. Exercise. Suppose a permutation a of $\mathbf{S}_n$ decomposes into disjoint cycles with lengths $n_1, \ldots, n_k$, $\sum n_i = n$. Evaluate $|N_{\mathbf{S}_n}(a)|$.

2.5.10. Exercise. Verify the formula

$$d^{-1}(\beta)t_{ij}(\alpha)d(\beta) = \begin{cases} t_{in}(\alpha\beta) & \text{if } j = n, \\ t_{nj}\left(\dfrac{\alpha}{\beta}\right) & \text{if } i = n, \\ t_{ij}(\alpha) & \text{in all other cases.} \end{cases}$$

Using this convince yourself that in Example 2.2.2(IV) the conclusion is valid with $r = 0$.

2.5.11. Exercise. The normalizers of the elements of a conjugacy class, themselves form a class of conjugate subgroups.

2.5.12. Exercise. In the group of matrices over the field $\mathbf{Q}$ of the form $\left(\begin{smallmatrix} \alpha & \beta \\ 0 & 1 \end{smallmatrix}\right)$, $\alpha \neq 0$, find a subgroup that is conjugate to a proper subgroup of itself.

2.5.13. Exercise. If A is a subgroup of finite index in a group G, then the intersection

$$N = \bigcap_{x \in G} A^x$$

is a normal subgroup of finite index in G. (Hint. Use Theorem 2.5.6 and Exercise 2.4.8.)

§3. The Center. The Commutator Subgroup

We have seen that groups in general have properties strikingly different from those of the various groups of numbers. In large part the differences are due to noncommutativity. The center and the commutator subgroup give a measure of the departure from commutativity: the bigger the center and the smaller the commutator subgroup of a group, the nearer the group is to being abelian.

3.1. The Center

Let M be a subset and H a subgroup of a group G. Recall that we defined the normalizer of M in H to be the set of all elements h of H commuting with M as a whole, i.e. such that $hM = Mh$. If instead we take the set of all elements of H which commute with each element of M, i.e. the set

$$C_H(M) = \{x \mid x \in H, m^x = m \text{ for all } m \in M\},$$

then we have what is called the *centralizer* of the set M in the subgroup H. It is not difficult to verify that $C_H(M)$ is a normal subgroup of $N_H(M)$. If M contains just one element then of course its normalizer and centralizer in H are the same. If there is no indication of the subgroup in which the centralizer is taken, then it is to be understood to be in the whole group G.

The centralizer of the whole group G is called its *center* and is denoted by $C(G)$. Obviously a group is abelian precisely when it is its own center. It is clear that the identity is always in the center. If a group contains no other central elements it is said to have *trivial center* or sometimes even to be *centerless*. Note further that every subgroup of the center is normal in the whole group.

To illustrate we calculate the centers of some of the groups of Examples (I)–(IV) of §1.

3.1.1. EXAMPLES. (I), (II). Since the additive and multiplicative group of a field are abelian, they coincide with their centers.

(III). Obviously the groups S_2 and A_3 are abelian and so coincide with their centers. The situation for other n is as follows. Any nonidentity permutation in S_n in its disjoint cycle decomposition has the form $(ij \cdots)(\cdots) \cdots$. It can be verified immediately that it does not commute with the transposition (jk) (here $n \geq 3$), nor with the 3-cycle (jkl) (here $n \geq 4$), where distinct letters denote distinct permuted objects. Hence the groups S_n for $n \geq 3$, and the groups A_n for $n \geq 4$, have trivial center.

(IV). Let K be a field. The centers of the groups $GL_n(K)$, $SL_n(K)$ consist of the scalar matrices contained in them. For obviously a scalar matrix commutes with any other matrix. Conversely if a is a central element of either of these groups then it commutes in particular with all the transvections $t_{ij}(1) = t_{ij}$, i.e. $at_{ij} = t_{ij}a$. It follows easily from this that

$$a_{ij} = 0, \qquad a_{ii} = a_{jj} \quad \text{for } i \neq j,$$

i.e. a is a scalar matrix. For this and similar calculations we recommend that the reader write each matrix x in the form $\sum x_{rs}e_{rs}$ and use the following easily reconstructed multiplication table for the matrices e_{ij}:

$$e_{ij}e_{rs} = \begin{cases} e_{is} & \text{for } j = r, \\ 0 & \text{for } j \neq r. \end{cases} \tag{1}$$

Similar reasoning shows that the center of the group $T_n(K)$, $|K| \neq 2$, consists of all nonzero scalar matrices, while in $UT_n^m(K)$ the central matrices other than e are just those that differ from e only in entries in the $m \times m$ submatrix in their upper right-hand corner. In particular

$$C(UT_n(K)) = UT_n^{n-1}(K) = \{t_{1n}(\alpha) \,|\, \alpha \in K\}. \tag{2}$$

The diagonal group $D_n(K)$ is abelian and so is its own center.

3.1.2. Exercise. Find the centralizer of a diagonal matrix in $GL_n(K)$.

3.1.3. Exercise. The centralizer of a normal subgroup is itself normal.

3.1.4. Exercise. If H is a finite normal subgroup of a group G, then the index of its centralizer is finite. (Solution:

$$|G : C_G(H)| = \left| G : \bigcap_{x \in H} C_G(x) \right| \leq \prod_{x \in H} |G : C_G(x)| < |H|^{|H|}.)$$

3.1.5. Exercise. The group of matrices $\left(\begin{smallmatrix} \alpha & \beta \\ 0 & 1 \end{smallmatrix} \right)$, $\alpha \neq 0$, over a field other than $GF(2)$, has trivial center.

3.1.6. Exercise. $|C(GL_n(q))| = q - 1$; $|C(SL_n(q))| = $ h.c.f.$(n, q - 1)$.

3.1.7. Exercise. The center of a direct (Cartesian) product is the direct (Cartesian) product of the centers of the factors.

3.2. The Commutator Subgroup

Obviously two elements a, b of a group G commute if and only if $a^{-1}b^{-1}ab = e$. The left-hand side of this equation is called the *commutator* of a, b in that order, and is written $[a, b]$. The subgroup of G generated by all commutators of pairs of elements of G, is called the *commutator subgroup* or *derived subgroup* of G. Clearly a group is abelian if and only if its commutator subgroup is trivial; generally speaking the commutator subgroup provides in some sense a measure of departure from commutativity.

More generally if L, M are subsets of a group G then their *mutual commutator subgroup* is the subgroup

$$[L, M] = \langle [a, b] \mid a \in L, b \in M \rangle.$$

Since

$$[a, b]^x = [a^x, b^x],$$

we have that the mutual commutator subgroup of a pair of normal subgroups is again normal. In particular the commutator subgroup $[G, G]$ of G is normal in G. Taking the commutator subgroup of the commutator subgroup, and so on, we obtain a descending chain of normal subgroups

$$G \geq G' \geq G'' \geq \cdots,$$

which is called the *derived series* of the group G.

We note the following useful and easily verified commutator identities:

$$[a, b]^{-1} = [b, a]; \qquad [ab, c] = [a, c]^b[b, c]; \qquad [a^{-1}, b] = [b, a]^{a^{-1}}. \quad (3)$$

Next, as usual, we use the groups of Examples (I)–(IV), §1, to illustrate our new concept.

3.2.1. EXAMPLES. (I), (II). Since the additive and multiplicative groups of a field are abelian, their commutator subgroups are trivial.

(III). Obviously $[\mathbf{S}_2, \mathbf{S}_2] = 1$, $[\mathbf{A}_3, \mathbf{A}_3] = 1$. Further

$$[\mathbf{A}_4, \mathbf{A}_4] = \{1, (12)(34), (13)(24), (14)(23)\}. \quad (4)$$

To see this observe first that the right-hand side is a subgroup (called Klein's 4-group). This subgroup is normal in $\mathbf{A}_4$ since it contains all permutations with disjoint cycle decomposition of the form $(**)(**)$, and we know that this form is preserved by conjugation. Since $\mathbf{A}_n$ is generated by all 3-cycles, the inclusion $\subseteq$ will follow from the identities (3) and the normality of the right-hand side of (4) once we have shown that all commutators of 3-cycles lie in the right-hand side of (4). But this is immediate from the inequalities

$$[(ijk), (ijl)] = (ij)(kl); \qquad [(ijk), (ilj)] = (il)(jk),$$

which can be checked routinely. (Here distinct letters denote distinct permuted objects.) These same equalities yield the reverse inclusion.

Finally,

$$[\mathbf{S}_n, \mathbf{S}_n] = \mathbf{A}_n \quad \text{for all } n, \tag{5}$$

$$[\mathbf{A}_n, \mathbf{A}_n] = \mathbf{A}_n \quad \text{for } n \geq 5. \tag{6}$$

For the commutator of two permutations from $\mathbf{S}_n$ is even and therefore lies in $\mathbf{A}_n$. On the other hand

$$(ijk) = [(ik), (ij)] = [(ikl), (ijm)],$$

where again different letters denote different objects. Since $\mathbf{A}_n$ is generated by 3-cycles the desired conclusions follow.

(IV). Let K be a field. Then

$$[\mathbf{GL}_n(K), \mathbf{GL}_n(K)] = \mathbf{SL}_n(K), \tag{7}$$

$$[\mathbf{SL}_n(K), \mathbf{SL}_n(K)] = \mathbf{SL}_n(K), \tag{8}$$

always, except for $\mathbf{GL}_2(2)$ in (7) and $\mathbf{SL}_2(2)$, $\mathbf{SL}_2(3)$ in (8). (Note that of course $\mathbf{GL}_2(2) = \mathbf{SL}_2(2)$, so that there are in fact only two exceptions in all.) To see (7) and (8) observe first that the determinant of the commutator of two matrices is always 1, whence follows the inclusion of the left-hand side of (7) in the right-hand side. In the other direction we know that $\mathbf{SL}_n(K)$ is generated by the transvections. It is easily verified that

$$[t_{ik}(\alpha), t_{kj}(\beta)] = t_{ij}(\alpha\beta) \quad \text{for distinct } i, j, k, \tag{9}$$

$$[t_{ij}(\alpha), \mathrm{diag}(\beta_1, \ldots, \beta_n)] = t_{ij}\left(\alpha\left(\frac{\beta_j}{\beta_i} - 1\right)\right). \tag{10}$$

In view of (9) the reverse inclusions (of the right-hand sides of (7), (8) in the left-hand sides) follow for $n \geq 3$. Since for $|K| > 2$ we can choose $\beta_1 \neq \beta_2$ in (10), we deduce the equation (7) for all the groups except $\mathbf{GL}_2(2)$. Since for $|K| > 3$ we can find $\beta_1 \neq \beta_2$, $\beta_1\beta_2 = 1$, we infer (8) from (10) for all the groups except $\mathbf{SL}_2(2)$ and $\mathbf{SL}_2(3)$. That these remaining groups are indeed exceptions can be shown without too much difficulty.

In a similar manner, using (3), (4) of §2 together with (9), (10), we obtain that

$$[\mathbf{T}_n(K), \mathbf{T}_n(K)] = \mathbf{UT}_n(K) \quad \text{for } |K| > 2, \tag{11}$$

$$[\mathbf{UT}_n^r(K), \mathbf{UT}_n^s(K)] = \mathbf{UT}_n^{r+s}(K). \tag{12}$$

When $|K| = 2$ the first equation is false, but then $\mathbf{T}_n(K) = \mathbf{UT}_n(K)$ so that the second equation applies.

3.2.2. Exercise. Find the derived series of the groups $\mathbf{S}_3$, $\mathbf{S}_4$.

3.2.3. Exercise. Find the commutator subgroups of the groups $\mathbf{SL}_2(2)$, $\mathbf{SL}_2(3)$.

3.2.4. Exercise. The commutator subgroup of a direct product of groups is the direct product of the commutator subgroups of the factors. Does this statement remain true with "direct" replaced by "Cartesian"?

3.2.5. Exercise. Find the commutator subgroup of the group of matrices $\left(\begin{smallmatrix} \alpha & \beta \\ 0 & 1 \end{smallmatrix}\right)$, $\alpha \neq 0$, over a field, and show that every element of that commutator subgroup is a commutator.

We draw the reader's attention to the fact that, in general, the commutator subgroup does not consist solely of commutators. There are even finite groups in which certain products of pairs of commutators are not commutators; we shall now give an example of such a group, including a sketch of the proof.

3.2.6. EXAMPLE. Consider the set S of 10 symbols with the following multiplication table:

	1	α	β	γ	δ	κ	λ	μ	ν	0
1	1	α	β	γ	δ	κ	λ	μ	ν	0
α	α	0	κ	λ	μ	0	0	0	0	0
β	β	0	0	ν	0	0	0	0	0	0
γ	γ	0	0	0	0	0	0	0	0	0
δ	δ	0	0	0	0	0	0	0	0	0
κ	κ	0	0	0	0	0	0	0	0	0
λ	λ	0	0	0	0	0	0	0	0	0
μ	μ	0	0	0	0	0	0	0	0	0
ν	ν	0	0	0	0	0	0	0	0	0
0	0	0	0	0	0	0	0	0	0	0

It is easy to check that multiplication in S is associative, i.e. S is, as they say, a *semigroup*. Let K be the semigroup-ring of S over the field $\mathbf{GF}(2)$, i.e. the ring of all formal expressions

$$n_1 1 + n_2\alpha + n_3\beta + n_4\gamma + n_5\delta + n_6\kappa + n_7\lambda + n_8\mu + n_9\nu + n_{10}0,$$

with coefficients from $\mathbf{GF}(2)$ and with the natural addition and multiplication. Finally let G be the set of all matrices over K of the form

$$a = \begin{pmatrix} 1 & x & y \\ 0 & 1 & x \\ 0 & 0 & 1 \end{pmatrix}.$$

It is easy to see that G is a group of order 2^{20}, and that if

$$b = \begin{pmatrix} 1 & u & v \\ 0 & 1 & u \\ 0 & 0 & 1 \end{pmatrix}, \quad \text{then } [a, b] = \begin{pmatrix} 1 & 0 & xu - ux \\ 0 & 1 & 0 \\ 0 & 0 & 1 \end{pmatrix}. \tag{13}$$

We call $xu - ux$ the *ring commutator* of the elements x, u.

3.2.7. Exercise. The element $\mu + \nu$ is the sum of two ring commutators, but is itself not a ring commutator.

3.2.8. Exercise. Using (13) and the preceding exercise, find in G a product of two commutators which is not a commutator.

We end this section, and the chapter, with two lemmas which will be useful to us somewhat later on (in Chapter 6).

3.2.9. Lemma. *Let G be an arbitrary group and let A, B be subgroups. Write $H = \langle A, B \rangle$. Then:*

$$[A, B], A[A, B], B[A, B] \text{ are normal in } H; \tag{14}$$

$$A[A, B] \cdot B[A, B] = H; \tag{15}$$

$$[A, B] = \langle C^{AB} \rangle, \tag{16}$$

where $C = \{[a_i, b_j] \mid i \in I, j \in J\}$, and the a_i, $i \in I$, and b_j, $j \in J$, are generators for A, B respectively.

PROOF. The second of the identities (3) shows that the subgroup A normalizes $[A, B]$, whence (14) and (15). For (16), observe that $C \subseteq [A, B] \trianglelefteq H$, giving that $[A, B] \geq \langle C^{AB} \rangle$. To get the reverse inclusion, we need to show that

$$[a_{i_1}^{\varepsilon_1} \cdots a_{i_m}^{\varepsilon_m}, b_{j_1}^{\delta_1} \cdots b_{j_n}^{\delta_n}] \in \langle C^{AB} \rangle, \quad \text{where here } \varepsilon_i = \pm 1, \ \delta_j = \pm 1.$$

Now we have

$$(a_{i_1}^{\varepsilon_1} \cdots a_{i_m}^{\varepsilon_m})^{b_j} = (a_{i_1}[a_{i_1}, b_j])^{\varepsilon_1} \cdots (a_{i_m}[a_{i_m}, b_j])^{\varepsilon_m}$$

$$= a_{i_1}^{\varepsilon_1} \cdots a_{i_m}^{\varepsilon_m} f, \qquad f \in \langle C^A \rangle;$$

that is

$$[a, b_j] \in \langle C^A \rangle, \quad \text{where } a = a_{i_1}^{\varepsilon_1} \cdots a_{i_m}^{\varepsilon_m}.$$

Hence

$$(b_{j_1}^{\delta_1} \cdots b_{j_n}^{\delta_n})^a = (b_{j_1}[b_{j_1}, a])^{\delta_1} \cdots (b_{j_n}[b_{j_n}, a])^{\delta_n}$$

$$= b_{j_1}^{\delta_1} \cdots b_{j_n}^{\delta_n} d, \qquad d \in \langle C^{AB} \rangle,$$

giving the desired conclusion. (Note that $\langle \langle C^A \rangle^B \rangle = \langle C^{AB} \rangle$.)

Let $a_1, a_2, \ldots,$ be elements, and $A_1, A_2, \ldots,$ be subgroups, of some group. For $n \geq 2$, we define inductively *left-normed commutators* by

$$[a_1, \ldots, a_{n+1}] = [[a_1, \ldots, a_n], a_{n+1}],$$

$$[A_1, \ldots, A_{n+1}] = [[A_1, \ldots, A_n], A_{n+1}].$$

3.2.10. Lemma ("Three Commutator Lemma"). *Let A, B, C be subgroups of some group, and let H be a normal subgroup. If two of the mutual* *commutator subgroups*

$$[A, B, C], [B, C, A], [C, A, B]$$

lie in H, then the third also lies in H.

PROOF. This is immediate from the following pretty identity ("Witt's identity"):

$$[a, b^{-1}, c]^b [b, c^{-1}, a]^c [c, a^{-1}, b]^a = 1. \tag{17}$$

To verify this simply expand the left-hand side.

3.2.11. Exercise. If A, B, C are normal subgroups of a group, then $[AB, C] = [A, C][B, C]$.

Homomorphisms 2

§4. Homomorphisms and Factors

4.1. Definitions

As defined at the beginning of Chapter 1, a map from one group to another is an isomorphism if it is one-to-one and onto, and preserves the operation. If the first two requirements are dropped we arrive at the concept of homomorphism: a map ϕ from a group G to a group G^* is said to be *homomorphic* or a *homomorphism* if $(ab)^\phi = a^\phi b^\phi$ for all a, b in G. The subgroup G^ϕ of G^* is the *homomorphic image* of G under ϕ.

4.1.1. EXAMPLES. (I). The map $\mathbf{Z} \to \mathbf{Z}_n$ sending each integer to its residue class modulo n, is a homomorphism. Let p be prime, and ε_n be a primitive p^n-th root of unity in the field of complex numbers such that $\varepsilon_{n+1}^p = \varepsilon_n$, $n = 1, 2, \ldots$. The map $\mathbf{Q}_p \to \mathbf{C}_{p^\infty}$, sending the rational number m/p^n to the complex number ε_n^m, is well-defined and homomorphic.

(II). The map $\mathbf{R}^* \to \mathbf{Z}^*$ which assigns to each real number its sign ± 1, is a homomorphism.

(III). The map $\mathbf{S}_n \to \mathbf{Z}^*$ assigning to each permutation its sign (± 1 according as it is even or odd), is a homomorphism.

(IV). The following are homomorphisms: the map $\mathbf{GL}_n(K) \to K^*$ associating matrices with their determinants; the map $\mathbf{T}_n(K) \to \mathbf{D}_n(K)$ sending each triangular matrix to the diagonal matrix with the same diagonal; the map $\mathbf{UT}_n^m(K) \to K \oplus \cdots \oplus K$ ($n - m$ times), sending a matrix x to the $(n - m)$-tuple $(x_{1,m+1}, x_{2,m+2}, \ldots, x_{n-m,n})$.

We see that after application of a homomorphism a group operation may lose certain of its properties: thus a nonabelian group may become abelian, a torsion-free group may become periodic, and so on, although the most "striking" properties—finiteness, commutativity, etc.—are preserved under epimorphisms (i.e. homomorphisms onto). The homomorphic image of a group G may be likened to the impression of the more essential features of G gained by an observer (whose point of view corresponds to the homomorphism). Being less detailed, this impression fails to characterize the original reality; however, in the first place this is precisely why it is easier to study, and secondly the impressions of several observers at various observation-posts may when taken together yield the desired information about G. For instance the knowledge that the groups $\mathbf{Z}_1, \mathbf{Z}_2, \ldots$ are all homomorphic images of a group G, allows us to assert that G is infinite, though knowing that any one of them is a homomorphic image does not by itself suffice for the assertion.

4.1.2. Exercise. The epimorphic image of a (normal) subgroup is a (normal) subgroup. The restriction of a homomorphism to a subgroup is a homomorphism of the subgroup.

Let ϕ be a homomorphism from a group G to a group G^*. The set of all elements of G mapped by ϕ to the identity, is called the *kernel* of ϕ, denoted by Ker ϕ. The kernel of a homomorphism is a normal subgroup. To see this write Ker $\phi = H$. Then since

$$(HH)^\phi = H^\phi H^\phi = 1, \qquad (H^{-1})^\phi = (H^\phi)^{-1} = 1, \qquad (H^G)^\phi = (H^\phi)^{G\phi} = 1,$$

it follows that

$$HH \subseteq H, \qquad H^{-1} \subseteq H, \qquad H^G \subseteq H,$$

so that $H \lhd G$. It is easy to see that the kernel of a homomorphism ϕ is trivial (Ker $\phi = 1$) if and only if ϕ is a monomorphism (i.e. a one-to-one homomorphism). Reverting to our simile we may say that the kernel of a homomorphism measures the clarity with which our observer views the group: the larger the kernel the less detail there is in his impression. The vaguest impression, namely that where only the fact that the group has an identity can be discerned, is afforded by the homomorphism with kernel the whole group.

It turns out that the set of kernels of homomorphisms and the set of normal subgroups of a group are one and the same; that is, every normal subgroup is the kernel of some homomorphism. The standard way of constructing a homomorphism with a given normal subgroup as kernel is as follows. Let G be a group, H the normal subgroup and G/H the set of cosets of H in G (there is no need to specify whether the cosets are left or right—since H is normal they coincide). It is easily verified that $aH \cdot bH = abH$, i.e. that the set G/H is closed under elementwise multiplication of its cosets. It is also easy to see that G/H is in fact a group with respect to coset multiplication; it is called the *quotient group* or *factor group* of the group G by the normal subgroup H. The identity element of the group G/H is the coset H, and the inverse of the coset aH is the coset $a^{-1}H$. We stress the distinction that, in the group G the residue class aH is a subset of elements, while in the group G/H it is a single element, not a subset. It will be immediately clear that the map $\phi: G \to G/H$ defined by the rule $g^{\phi} = gH$ is a homomorphism. Such homomorphisms are termed *natural*. They solve our problem: the kernel of the natural homomorphism $G \to G/H$ is just H.

4.1.3. Exercise. A factor group of a cyclic group is cyclic.

4.1.4. Exercise. The factor group G/H is abelian if and only if H contains the derived group $[G, G]$.

A different approach to the concept of factor group is given by

4.1.5. Exercise. An equivalence relation $\sim$ on a group G is called a *congruence* if we always have

$$a \sim b, a' \sim b' \Rightarrow aa' \sim bb'.$$

The product of two congruence classes of elements is again a congruence class. The set $G/\sim$ of all congruence classes is a group under multiplication of classes, called the *factor group with respect to the congruence* $\sim$. This approach has the virtue that, as is easily seen, it works for any algebraic system (i.e. set with operations).

4.1.6. Exercise. The congruences on a group G are in one-to-one correspondence with the normal subgroups of G. To be more explicit, if $H \trianglelefteq G$ and if $\sim$ is defined by

$$a \sim b \Leftrightarrow a^{-1}b \in H,$$

then the relation $\sim$ is a congruence and the cosets of H in G are just the congruence classes. Conversely, given a congruence $\sim$ on G, the set H of elements congruent to the identity turns out to be a normal subgroup of G, and the congruence classes are just the cosets of H in G. Thus the factor group with respect to the congruence is the same as the factor group of G by H. Although, as mentioned above, the concept of quotient with respect to a congruence makes sense for other algebraic systems, there is often no reasonable analogue of the concept of normal subgroup. Hence while for groups the two approaches lead to the same concept, the approach via congruence has the wider scope outside groups.

4.2. Homomorphism Theorems

For G a group and H a subgroup we denote by $L(G, H)$ the set of all subgroups of G containing H. In particular $L(G, 1)$ is the set of all subgroups of G. The following theorem is about correspondences of subgroups under homomorphisms.

4.2.1. Theorem. *The natural homomorphism* $\phi: G \to G/H$ *induces a one-to-one correspondence* $\psi: L(G, H) \to L(G/H, 1)$. *Subgroups* A, B *belonging to* $L(G, H)$ *are conjugate in* G *if and only if their images* $A\psi$, $B\psi$ *are conjugate in* G/H. *In particular* A *is normal in* G *if and only if* $A\psi$ *is normal in* G/H. *If* $A \leq B$ *then* $|B:A| = |B\psi : A\psi|$.

PROOF. We first prove that ψ is one-to-one. For this it suffices to show that if $a \in A$, $a \notin B$, then $a^\phi \in A^\phi$, $a^\phi \notin B^\phi$. Thus suppose $a^\phi \in B^\phi$, say $a^\phi = b^\phi$, $b \in B$. Then $b^{-1}a \in H \leq B$ whence $a \in B$, contrary to assumption. The map ψ is *onto* since the preimage under ψ of a given subgroup $\bar{A}$ of G/H is just the complete inverse image of $\bar{A}$ under ϕ. A similarly direct check shows that

$$B = A^x \Leftrightarrow B^\phi = (A^\phi)^{x^\phi}.$$

Finally, if $A \leq B$ then from

$$x^{-1}y \in A \Leftrightarrow (xH)^{-1}(yH) \in A/H,$$

it follows that the cosets of A in B are in one-to-one correspondence with the cosets of A/H in B/H. This completes the proof.

It turns out that every homomorphism is in essence a natural homomorphism, or, more precisely, every homomorphism is the composite of a natural homomorphism and a monomorphism. We shall now also discover that: "Observers at observation-posts corresponding to homomorphisms with the same kernel, receive similar (i.e. isomorphic) impressions."

4.2.2. Theorem. *If* $\phi: G \to K$ *is a homomorphism with kernel* H, *then* $G^\phi \simeq G/H$. *Further,* ϕ *is the composite of the natural homomorphism* $\varepsilon: G \to G/H$, *and a certain monomorphism* $\tau: G/H \to K$, *defined by* $(xH)^\tau = x^\phi$.

PROOF. The map τ is well-defined since $xH = yH$ implies $x^\phi = y^\phi$. It is also one-to-one since if $x^\phi = y^\phi$ then $x^{-1}y \in H$. Finally, τ preserves multiplication since

$$(xH \cdot yH)^\tau = (xyH)^\tau = (xy)^\phi = x^\phi y^\phi = (xH)^\tau (yH)^\tau.$$

Hence τ is a monomorphism. Since it is clear that $\phi = \varepsilon\tau$ the theorem is proved.

To illustrate we apply this theorem to Examples 4.1.1. If we compute the kernels of the homomorphisms in those examples we obtain the following isomorphisms:

$$\mathbf{Z}/\langle n \rangle \simeq \mathbf{Z}_n; \tag{1}$$

$$\mathbf{Q}_p/\mathbf{Z} \simeq \mathbf{C}_{p^\infty}; \tag{2}$$

$$\mathbf{S}_n/\mathbf{A}_n \simeq \mathbf{Z}_2; \tag{3}$$

$$\mathbf{GL}_n(K)/\mathbf{SL}_n(K) \simeq K^*; \tag{4}$$

$$\mathbf{T}_n(K)/\mathbf{UT}_n(K) \simeq \mathbf{D}_n(K) \simeq K^* \times \cdots \times K^* \ (n \ \text{times}); \tag{5}$$

$$\mathbf{UT}_n^m(K)/\mathbf{UT}_n^{m+1}(K) \simeq K \oplus \cdots \oplus K \ (n - m \ \text{times}). \tag{6}$$

Using Theorem 4.2.2 we can elicit a further fact about the correspondence between $L(G, H)$ and $L(G/H, 1)$ of Theorem 4.2.1, namely that corresponding normal subgroups yield isomorphic factor groups of G:

4.2.3. Theorem. *If* $H \lhd G$, $A \lhd G$ *and* $H \le A$, *then*

$$(G/H)/(A/H) \simeq G/A.$$

PROOF. Define $\phi : G/H \to G/A$, by

$$(xH)^\phi = xA, \qquad x \in G.$$

The map ϕ is well-defined since from $xH = yH$ follows $x^{-1}y \in H \le A$, so that $xA = yA$. That ϕ is onto is clear. Finally ϕ preserves multiplication since

$$(xH \cdot yH)^\phi = (xyH)^\phi = xyA = xA \cdot yA.$$

Hence ϕ is an epimorphism. It is obvious that $\operatorname{Ker} \phi = A/H$, so that $A/H \lhd G/H$, and the theorem then follows from Theorem 4.2.2.

4.2.4. Theorem. *If* $A \lhd B \le G$, $H \lhd G$, *then*

$$BH/AH \simeq B/A(B \cap H).$$

In particular,

$$BH/H \simeq B/B \cap H.$$

PROOF. Let θ be the restriction to B of the natural homomorphism $\phi : G \to G/H$. Then $\operatorname{Ker} \theta = B \cap H$. Since $A^\theta = A^\phi$, $B^\theta = B^\phi$, the complete inverse images under θ of A^θ and B^θ will be $A(B \cap H)$ and B. Since $A \lhd B$, we have $A^\theta \lhd B^\theta$. It follows from Theorem 4.2.1 that $A(B \cap H) \lhd B$, and from Theorem 4.2.3 that

$$B/A(B \cap H) \simeq B^\theta/A^\theta = (BH/H)/(AH/H) \simeq BH/AH.$$

4.2.5. Exercise. If $G = A \times B$, then $G/A \simeq B$.

4.2.6. Exercise. Let H be the subgroup of $\mathbf{C}^*$ consisting of all complex numbers of modulus 1, and let $\mathbf{R}^{*+}$ denote the multiplication group of positive reals. Then $\mathbf{C}^*/H \simeq \mathbf{R}^{*+}$.

Without doubt the concepts of group, subgroup, normal subgroup, homomorphism and factor group were familiar to the reader from his course in general algebra. However in a book entitled "Fundamentals of the theory of groups," it was surely not out of place to mention them again. It *would* however be superfluous to mention the analogous concepts of subring, ideal, ring homomorphism, and factor ring. There is an interesting connexion between these and other concepts on the one hand, and group theory on the other, arising from the study of groups of matrices over rings.

4.2.7. Exercise. Let K be an associative ring with a multiplicative identity, and G the group of matrices of degree n over K. If I is an ideal in K, then

$$G_I = \{e + a \mid e + a \in G, a_{ij} \in I\}$$

is a normal subgroup of G. It is called the *principal congruence subgroup modulo I*, and its supergroups are the *congruence subgroups* of the group G. Every ring homomorphism $\phi: K \to K_1$, induces a group homomorphism $\phi_G: G \to G_1$, defined by

$$(g_{ij})^{\phi_G} = (g_{ij}^{\phi}), \qquad g \in G.$$

The kernel $\operatorname{Ker} \phi_G = G_{\operatorname{Ker} \phi}$. For example the homomorphism $\mathbf{Z} \to \mathbf{Z}_m$ induces a homomorphism $\mathbf{SL}_n(\mathbf{Z}) \to \mathbf{SL}_n(\mathbf{Z}_m)$, whose kernel (the mth principal congruence subgroup) is denoted by $\Gamma_n(m)$. Each $\Gamma_n(m)$ has finite index in $\mathbf{SL}_n(\mathbf{Z})$.

In connexion with this last exercise we mention the important *congruence problem* for matrix groups G: Does every subgroup of finite index in G contain G_I for some ideal I of finite index? We shall give on p. 163 one of the earliest known examples for which the answer is in the affirmative.

J. L. Mennicke [Finite factor groups of the unimodular group, Ann. Math. **81** (1965), 31–37] and independently H. Bass, M. Lazard and J.-P. Serre [Sous-groupes d'indice fini dans $\mathbf{SL}_n(\mathbf{Z})$, Bull. Amer. Math. Soc. **70** (1964), 385–392] solved the congruence problem affirmatively for the group $\mathbf{SL}_n(\mathbf{Z})$, $n \geq 3$ (in the case $n = 2$ it has long been known to have a negative solution). Mennicke's paper is accessible to second-year students.

4.3. Subcartesian Products

Recall that the direct product

$$G = \prod_{i \in I}^{\times} G_i \tag{7}$$

consists of the functions $f: I \to \bigcup_{i \in I} G_i$, satisfying the two conditions:

$$f(i) \in G_i, \qquad i \in I;$$

$$|\operatorname{supp} f| < \infty.$$

It is easily verified that the subset

$$\hat{G}_i = \{f \mid f \in G, f(j) = e \text{ for } j \neq i\}$$

is a normal subgroup of G, isomorphic to the factor G_i under the map $f \to f(i)$. On the strength of this isomorphism $\hat{G}_i$ is often identified with G_i. With this identification, we have that $G_i \trianglelefteq G$, that the group G is generated by its subgroups G_i, and that each element $g \in G$ can be written uniquely as

$$g = g_{i_1} \cdots g_{i_m}, \tag{8}$$

where the subscripts $i_1, \ldots, i_m$ are distinct and $e \neq g_{i_j} \in G_{i_j}$, $j = 1, \ldots, m$.

This remark leads us to the definition of the internal direct product, as distinct from the external direct product defined in Chapter 1 and again above. Suppose that in a group G there is a set of normal subgroups G_i, such that each nontrivial element of G has a unique expression (8). We then say that the group G *decomposes as the direct product of its subgroups* G_i. Obviously a group G decomposes as a direct product of its subgroups G_i if and only if it is isomorphic to the external direct product of the (abstract) groups G_i. Hence the notation (7) will be used also to indicate that G is the internal direct product of its subgroups G_i. In additive terminology we speak of the decomposition of a group as a direct sum and write as before

$$G = G_1 \oplus \cdots \oplus G_m, \qquad G = \bigoplus_{i \in I} G_i.$$

4.3.1. EXAMPLES. (I). The additive group of the field $\mathbf{C}$ decomposes as the direct sum of the group of additive reals and the group of purely imaginary numbers: $\mathbf{C} = \mathbf{R} \oplus i\mathbf{R}$.

(II). Obviously,

$$\mathbf{Q}^* = \langle -1 \rangle \times \prod_p{}^\times \langle p \rangle.$$

(III). Klein's four-group decomposes as the direct product of two subgroups of order 2:

$$\{1, (12)(34), (13)(24), (14)(23)\} = \langle (12)(34) \rangle \times \langle (13)(24) \rangle.$$

(IV). Obviously,

$$\mathbf{D}_n(K) = G_1 \times \cdots \times G_n,$$

where G_i is the subgroup of $\mathbf{D}_n(K)$ consisting of those diagonal matrices whose diagonal entries, except possibly for the ith, are 1.

4.3.2. Exercise. A group decomposes as a direct product of normal subgroups G_i if and only if it is generated by them, and relations of the form $g_{i_1} \cdots g_{i_m} = e$, where $g_i \in G_i$, and the subscripts i_j are distinct, imply $g_{i_1} = \cdots = g_{i_m} = e$.

4.3.3. Exercise. A group decomposes as a direct product of normal subgroups G_i if and only if it is generated by them, and

$$G_i \cap \langle G_j \mid j \neq i \rangle = 1 \quad \text{for all } i.$$

4.3.4. Exercise. A cyclic group of order mn where m, n are coprime, decomposes into the direct product of its subgroups of orders m, n.

4.3.5. Exercise. Let p be a prime. The cyclic group of order p^n is indecomposable (as a direct product of proper subgroups).

4.3.6. Exercise. The additive group of the field $\mathbf{Q}$ is directly indecomposable.

A subgroup A of the direct product (7) is said to be a *subdirect product* of the groups G_i, if the projection of A on each factor G_i is the whole of G_i. It should be emphasized that a subdirect product is not uniquely determined by the factors. Obviously every subgroup of a direct product is a subdirect product of its projections. A subdirect product need not of course be directly decomposable: the diagonal subgroup D of the direct square $G \times G$, defined by $D = \{(g, g) \mid g \in G\}$, will serve as counterexample if we take G to be cyclic of prime order, for instance.

4.3.7. Exercise. Let $G = G_1 \times G_2$, A be a subgroup of G, and A_i the projection of A on G_i, $i = 1, 2$. Prove that $A = A_1 \times A_2$ if and only if $A_i = G_i \cap A$, $i = 1, 2$.

4.3.8. Exercise. Let $G = G_1 \times G_2$ where G_1, G_2 are finite groups of coprime orders. Every subgroup $A \le G$ is the direct product of its projections on the factors G_1, G_2.

If G is a subdirect product of finite groups then clearly any finite set of elements of G is contained in a finite normal subgroup. Groups with the latter property are usually called *locally normal*. As the group $\mathbf{C}_{p^\infty}$ shows, the class of locally normal groups is larger than that of subdirect products of finite groups. There are however close links between these classes [P. Hall, Periodic FC-groups, J. London Math. Soc. **34** (1959), 289–304; Ju. M. Gorčakov, On locally normal groups, Matem. sb. **67** (1965), 244–254].

Analogously to subdirect products one defines subcartesian products: A subgroup A of the Cartesian product

$$G = \prod_{i \in I} G_i$$

is called a *subcartesian product* of the groups G_i if the projection of A on each factor G_i is the whole of G_i. Clearly a subdirect product of groups G_i will also be a subcartesian product of those groups.

4.3.9. Theorem (Remak). *Suppose that in a group G we are given a family of normal subgroups H_i, $i \in I$, with intersection H. Then the factor group G/H is isomorphic to some subcartesian product of the factor groups G/H_i.*

PROOF. Consider the map

$$\phi: G \to \prod_{i \in I} (G/H_i),$$

which sends each $g \in G$ to the function f given by $f(i) = gH_i$. It is easily verified that ϕ is a homomorphism with kernel H. The result then follows by one of the homomorphism theorems (Theorem 4.2.2).

4.3.10. EXAMPLES. (I). The group $\mathbf{Q}/\mathbf{Z}$ is isomorphic to a subcartesian product of the groups $\mathbf{Q}/\mathbf{Q}_p$ where p ranges over the set of primes.

(II). The group $\mathbf{C}^*$ is isomorphic to a subdirect product of the groups $\mathbf{C}^*/\mathbf{C}_{p^\infty}$ for any two distinct primes p.

(III). Let M be a set partitioned into (pairwise nonintersecting) subsets M_i, $i \in I$. Let $G \leq \mathbf{S}(M)$ be such that every element of G maps M_i onto itself for all i. Then G is isomorphic to a subcartesian product of subgroups G_i of the $\mathbf{S}(M_i)$, $i \in I$.

(IV). The group $\mathbf{GL}_n(\mathbf{Z})$ is isomorphic to a subgroup of the Cartesian product of the finite groups $\mathbf{GL}_n(\mathbf{Z}_m)$, $m = 1, 2, \ldots$.

4.3.11. Exercise. If N_i, $i \in I$, are normal subgroups of a group G such that G/N_i is abelian for all i, then so also is $G/\bigcap_{i \in I} N_i$ abelian.

4.4. Subnormal Series

Each normal subgroup H of a group G determines a chain $1 \leq H \leq G$. This is a particular instance of the more general situation of a finite chain

$$1 = G_0 \leq G_1 \leq \cdots \leq G_n = G \tag{9}$$

of normal subgroups of G, known as a *normal series* for G. For example any finite chain of subgroups of an abelian group is a normal series. The chain (9) is called a *subnormal series* for G if it satisfies the weaker requirement that each member be normal in its successor, i.e. $G_i \trianglelefteq G_{i+1}$ for $i = 0, 1, 2, \ldots, n-1$. A subgroup H which occurs in some subnormal series for G is said to be *subnormal* in G, and we write $H \triangleleft\triangleleft G$. The factor groups G_{i+1}/G_i are called *factors* of the series (9), and the number n its *length*. (More generally a *factor* of a group G is any factor group B/A, where A, B are subgroups of G with $A \trianglelefteq B$.) Sometimes series are written in descending order from G to 1 instead of ascending order from 1 to G. Knowledge of the factors of some subnormal series of a group furthers our knowledge of the group itself.

We say that a group G is an *extension* of a group A by a group B if there is a normal subgroup H of G such that $H \simeq A$, $G/H \simeq B$. In this sense the factors of a subnormal series are building blocks from which by a succession of extensions one can construct the original group. We must emphasize however that the resulting group is not determined solely by its blocks, but depends also on the way they are placed on top of one another—for example there are two extensions of $\mathbf{Z}_2$ by $\mathbf{Z}_2$, namely $\mathbf{Z}_2 \oplus \mathbf{Z}_2$ and $\mathbf{Z}_4$. A group built of cyclic blocks is called polycyclic. More precisely, a subnormal series with cyclic factors is said to be *polycyclic*, and a group with such a series a *polycyclic group*.

4.4.1. EXAMPLES. (I). The factors of the series $1 < \langle 2 \rangle < \mathbf{Z} < \mathbf{Q}_p$ are isomorphic to $\langle 2 \rangle$, $\mathbf{Z}_2$, $\mathbf{C}_{p^\infty}$ respectively.

(II). The factors of the series $1 < \mathbf{C}_p < \mathbf{C}_{p^2} < \cdots < \mathbf{C}_{p^n}$ are all cyclic of order p.

(III). The group $\mathbf{A}_4$ possesses a polycyclic series

$$1 < \{1, (12)(34)\} < \{1, (12)(34), (13)(24), (14)(23)\} < \mathbf{A}_4, \tag{10}$$

with factors of orders 2, 2, 3. This series is not normal (since the subgroup $\langle (12)(34) \rangle$ is not normal in $\mathbf{A}_4$) although it is subnormal.

(IV). The group $\mathbf{T}_n(K)$ has a normal series

$$\mathbf{T}_n(K) \ge \mathbf{UT}_n^1(K) \ge \mathbf{UT}_n^2(K) \ge \cdots \ge \mathbf{UT}_n^n(K) = 1, \tag{11}$$

with factors $F_0, F_1, \ldots, F_{n-1}$, where

$$F_0 \simeq K^* \times \cdots \times K^* \quad (n \text{ times}),$$

$$F_m \simeq K \oplus \cdots \oplus K \quad (n-m \text{ times}) \quad \text{for } m \ge 1.$$

A (sub)normal series of a group induces a (sub)normal series of its subgroups and factor groups:

4.4.2. Theorem. *Let G be a group with a (sub)normal series (9). If $H \le G$ then by intersecting the series (9) with H we obtain a (sub)normal series for H:*

$$1 = H_0 \le H_1 \le \cdots \le H_n = H, \qquad H_i = G_i \cap H,$$

where each factor H_{i+1}/H_i is isomorphic to a subgroup of G_{i+1}/G_i. If $H \trianglelefteq G$ then by taking the images of the terms of (9) under the natural homomorphisn $G \to G/H$, we obtain a (sub)normal series for G/H:

$$1 = \hat{G}_0 \le \hat{G}_1 \le \cdots \le \hat{G}_n = G/H, \qquad \hat{G}_i = G_i H/H,$$

where each factor $\hat{G}_{i+1}/\hat{G}_i$ is a homomorphic image of the factor G_{i+1}/G_i.

PROOF. It is almost immediate that $H_i \trianglelefteq H_{i+1}$, $\hat{G}_i \trianglelefteq \hat{G}_{i+1}$, and in the normal series case that $H_i \trianglelefteq H$, $\hat{G}_i \trianglelefteq \hat{G}_n$. Further, using the homomorphism theorems we get:

$$H_{i+1}/H_i = H_{i+1}/H_{i+1} \cap G_i \simeq H_{i+1} G_i/G_i \le G_{i+1}/G_i,$$

$$\hat{G}_{i+1}/\hat{G}_i \simeq G_{i+1} H/G_i H \simeq G_{i+1}/G_i(G_{i+1} \cap H) \simeq (G_{i+1}/G_i)/(\cdots).$$

The theorem is proved.

4.4.3. Exercise. The subgroups and factor groups of a polycyclic group are polycyclic. The class of polycyclic groups is also closed under extensions.

Two subnormal series of a group are said to be *isomorphic* if they have the same length and there is a one-to-one correspondence between their factors such that corresponding factors are isomorphic. For example the two series

$$1 < C_2 < C_4 < C_{12},$$
$$1 < C_3 < C_6 < C_{12},$$

of the group C_{12}, are isomorphic. A series of a group is called a *refinement* of another series of the group if the terms of the second series all occur as terms of the first.

4.4.4. Theorem (Schreier). *Any two (sub)normal series of a group possess isomorphic (sub)normal refinements.*

PROOF. Suppose given two (sub)normal series of a group G:

$$1 = A_0 \leq A_1 \leq \cdots \leq A_m = G, \tag{12}$$
$$1 = B_0 \leq B_1 \leq \cdots \leq B_n = G. \tag{13}$$

Between each pair $A_i \leq A_{i+1}$ of consecutive terms of the first series we shall insert a chain of subgroups determined partly by the second series, and vice versa. The chain to be inserted between A_i and A_{i+1} is constructed as follows: Intersect the series (13) with A_{i+1} and multiply each member of the resulting series by A_i; the end result is a chain of subgroups beginning with A_i and ending with A_{i+1}—this is the chain to be inserted. We write it as

$$A_i = C_{i0} \leq C_{i1} \leq \cdots \leq C_{in} = A_{i+1}, \qquad C_{ij} = (A_{i+1} \cap B_j)A_i.$$

We construct analogous chains to be inserted in the second series:

$$B_j = D_{j0} \leq D_{j1} \leq \cdots \leq D_{jm} = B_{j+1}, \qquad D_{ji} = (B_{j+1} \cap A_i)B_j.$$

Obviously if the series (12), (13) are normal then the terms C_{ij}, D_{ji} of the inserted chains will also be normal in G. It only remains to prove that

$$C_{i,j+1}/C_{ij} \simeq D_{j,i+1}/D_{ji}.$$

Now by Theorem 4.2.4:

$$C_{i,j+1}/C_{ij} = (A_{i+1} \cap B_{j+1})A_i/(A_{i+1} \cap B_j)A_i$$
$$\simeq (A_{i+1} \cap B_{j+1})/(A_{i+1} \cap B_j)(A_i \cap B_{j+1});$$
$$D_{j,i+1}/D_{ji} = (B_{j+1} \cap A_{i+1})B_j/(B_{j+1} \cap A_i)B_j$$
$$\simeq (B_{j+1} \cap A_{i+1})/(B_{j+1} \cap A_i)(B_j \cap A_{i+1}).$$

This completes the proof.

From this theorem it follows in particular that *two (sub)normal series of a group are isomorphic if they have no repeated terms and cannot be refined except by repeating terms* (Jordan-Hölder theorem). Such a subnormal series is called a *composition series*.

4.4.5. Exercise. Let G be a polycyclic group. The number of infinite factors is the same for all polycyclic series of G. It is called the *polycyclic rank* of the group G. In the study of polycyclic groups it is often convenient to use induction on this number.

Observe that Theorem 4.4.2 and Schreier's theorem carry over without change to infinite series of the form

$$1 = G_0 \trianglelefteq G_1 \trianglelefteq \cdots \trianglelefteq G_n \trianglelefteq \cdots, \qquad G = \bigcup_{n=1}^{\infty} G_n,$$

and also to certain more complicated kinds of "infinite subnormal series." To avoid misunderstanding we note that in some textbooks subnormal and normal series are called instead normal and invariant series respectively. This is less apt since the terms of an invariant series must then be called normal subgroups, and the terms of a normal series subnormal subgroups, or else quite different words used—"accessible" and so on. In this book—as also, incidentally, in most of the group-theoretical literature—the words "invariant" and "accessible" remain unrequisitioned and will be used henceforth only in their nontechnical senses.

To conclude this section we mention the concept of residuality, a concept closely connected with that of homomorphism. In its most general sense this concept is defined as follows ([21], p. 75; we shall define it only in the context of groups, although it makes sense for arbitrary algebraic systems—see §24.1 of the appendix).

Let G be a group and ρ a relation between (or equivalently a predicate on families of) elements and sets of elements, defined on G and all its homomorphic images (for example: the binary relation of equality of elements; the binary relation "the element x belongs to the subgroup y"; the binary relation of conjugacy of elements; and so on). Let $\mathfrak{K}$ be a class of groups. We shall say that the group G is *residually in $\mathfrak{K}$ relative to ρ*, if for each family of elements and sets of elements of G not in the relation ρ, there is an epimorphism of G onto a group from the class $\mathfrak{K}$, such that the image of the family under this epimorphism is still not in the relation ρ. In the literature one meets up most often with residuality relative to equality of elements; in this case reference to the relation is usually omitted, so that one speaks of residuality *tout court*. By Remak's theorem (4.3.9) a group is residually in a class $\mathfrak{K}$ if and only if it is isomorphic to a subcartesian product of groups from $\mathfrak{K}$.

The property of being residually in the class of finite groups is called more briefly *residual finiteness*. Residual finiteness relative to a predicate ρ is conveniently denoted by $R\widetilde{\mathfrak{F}}\rho$; in particular letting ρ be in turn the predicates of: equality, conjugacy, belonging to a subgroup, belonging to a finitely generated subgroup, belonging to a fixed subgroup H, and so on, we obtain the properties (and classes) $R\widetilde{\mathfrak{F}}$, $R\widetilde{\mathfrak{F}}C$, $R\widetilde{\mathfrak{F}}\epsilon$, $R\widetilde{\mathfrak{F}}\epsilon_\omega$, $R\widetilde{\mathfrak{F}}\epsilon H$, and so on. The importance of these properties has been pointed out by A. I. Mal'cev [On homomorphisms onto finite groups, Uč. zap. Ivanovskogo ped. in-ta **18** (1958), 49–60]: the presence of any of them in a group implies the algorithmic decidability of the corresponding problem.

§5. Endomorphisms. Automorphisms

5.1. Definitions

A homomorphism of a group to itself is called an *endomorphism* of the group; a one-to-one, onto endomorphism is an *automorphism*. An endomorphic image is like the testimony to a person's character afforded by the contents of his pockets. The sets of all endomorphisms and of all automorphisms of a group G are denoted respectively by End G and Aut G. These sets are equipped with a multiplication, namely composition of endomorphisms. It is easily verified that then End G is a semigroup, while Aut G is even a group; in fact

$$\text{Aut } G \leq \mathbf{S}(G).$$

If G is abelian we can define addition on End G by setting

$$x(\phi + \psi) = x\phi + x\psi \quad \text{for } \phi, \psi \in \text{End } G, \qquad x \in G.$$

It is almost immediate that End G then becomes a ring. For nonabelian G we do not get a ring in this way, so that in general in group theory End plays a smaller role than Aut.

As has been remarked before (in Chapter 1), conjugation of a group G by an element a of G is an automorphism since it is one-to-one, onto and

$$(xy)^a = x^a y^a \quad \text{for } x, y \in G.$$

This automorphism, ι_a say, is called the *inner automorphism* of G, induced by the element a. It is immediate that for $a, b, x \in G$, $\phi \in$ Aut G,

$$(x^a)^b = x^{ab}, \qquad (x^a)^{a^{-1}} = x, \qquad x^{\phi^{-1}\iota_a\phi} = x^{\iota_{a^\phi}},$$

whence

$$\iota_a \iota_b = \iota_{ab}, \qquad \iota_a^{-1} = \iota_{a^{-1}}, \qquad \phi^{-1}\iota_a\phi = \iota_{a^\phi} \tag{1}$$

It follows from this that the set Inn G of all inner automorphisms of a group G is a normal subgroup of Aut G:

$$\text{Inn } G \trianglelefteq \text{Aut } G. \tag{2}$$

An automorphism which is not inner is called *outer*, and we shall call the group

$$\text{Out } G = \text{Aut } G/\text{Inn } G, \tag{3}$$

the *group of outer automorphisms*. Obviously for an abelian group Out coincides with Aut.

The first of the equations (1) shows that the map $G \to \text{Inn } G$ sending each g in G to the inner automorphism ι_g induced by it, is an epimorphism (homomorphism onto). The kernel of this epimorphism consists of those elements g satisfying

$$x^g = x, \qquad x \in G;$$

that is, the kernel is just the center $C(G)$ of the group G. Applying the appropriate homomorphism theorem we get that

$$\text{Inn } G \simeq G/C(G). \tag{4}$$

By way of illustration we shall describe the rings End G and the groups Aut G for some of the groups G of Examples (I)–(IV) of Chapter 1. We shall need the obvious equality

$$(\text{End } G)^* = \text{Aut } G.$$

5.1.1. EXAMPLES. (I). We have

$$\text{End } \mathbf{Z} \simeq \mathbf{Z}, \qquad \text{Aut } \mathbf{Z} \simeq \mathbf{Z}^* \simeq \mathbf{Z}_2, \tag{5}$$

$$\text{End } \mathbf{Z}_m \simeq \mathbf{Z}_m, \qquad \text{Aut } \mathbf{Z}_m \simeq \mathbf{Z}_m^*, \tag{6}$$

$$\text{End } \mathbf{Q} \simeq \mathbf{Q}, \qquad \text{Aut } \mathbf{Q} \simeq \mathbf{Q}^*. \tag{7}$$

In fact the three left-hand isomorphisms are defined as follows:

$$\phi \to 1\phi, \qquad \phi \to 1(\text{mod } m)\phi, \qquad \phi \to 1\phi.$$

We consider in detail only the last case. It is immediate that the map $\phi \to 1\phi$ defines a homomorphism. We show that the kernel is trivial, i.e. that if $1\phi = 0$, then $\phi = 0$. Let r, s be integers, $s \neq 0$. Since

$$0 = 1\phi = \left(s \cdot \frac{1}{s}\right)\phi = s\left(\frac{1}{s}\phi\right),$$

we have

$$\left(\frac{1}{s}\right)\phi = 0, \qquad \left(\frac{r}{s}\right)\phi = 0, \qquad \phi = 0.$$

Finally, for every $\alpha \in \mathbf{Q}$ there is a preimage $\phi \in \text{End } \mathbf{Q}$, namely the endomorphism $x \to \alpha x$ of the group $\mathbf{Q}$.

(II). We describe the ring End $\mathbf{C}_{p^\infty}$. Let ε_n, $n = 1, 2, \ldots$, be primitive complex p^n-th roots of 1 such that $\varepsilon_{n+1}^p = \varepsilon_n$. Obviously any endomorphism of the group $\mathbf{C}_{p^\infty}$ is completely determined by its action on $\varepsilon_1, \varepsilon_2, \ldots$. Let $\phi \in \mathrm{End}\, \mathbf{C}_{p^\infty}$, then

$$\varepsilon_n^\phi = \varepsilon_n^{k_n}, \qquad k_n \in \mathbf{Z}_{p^n}, \qquad n = 1, 2, \cdots.$$

Since $(\varepsilon_{n+1}^\phi)^p = \varepsilon_n^\phi$, it follows that k_{n+1} goes to k_n under the natural homomorphism $\mathbf{Z}_{p^{n+1}} \to \mathbf{Z}_{p^n}$ (defined by $x(\mathrm{mod}\, p^{n+1}) \to x(\mathrm{mod}\, p^n)$, for $x \in \mathbf{Z}$). Thus to each ϕ in $\mathrm{End}\, \mathbf{C}_{p^\infty}$ there corresponds a sequence $(k_1, k_2, \ldots)$, $k_n \in \mathbf{Z}_{p^n}$, satisfying the condition that k_n is the image of k_{n+1} under the natural homomorphism $\mathbf{Z}_{p^{n+1}} \to \mathbf{Z}_{p^n}$. It is immediate that the set $\mathbf{Z}_{p^\infty}$ of all such sequences is a ring under termwise addition and multiplication (the "inverse limit" of the $\mathbf{Z}_{p^n}$), and then our map $\mathrm{End}\, \mathbf{C}_{p^\infty} \to \mathbf{Z}_{p^\infty}$ can easily be seen to be a ring epimorphism. This gives the description of the ring $\mathrm{End}\, \mathbf{C}_{p^\infty}$ that we are looking for. The ring $\mathbf{Z}_{p^\infty}$ is called the *ring of p-adic integers*; its elements can be written in a natural way in the convenient form

$$\cdots a_n \cdots a_1 a_0 = \sum_{n=0}^{\infty} a_n p^n, \qquad 0 \le a_n < p,$$

and in this form they can be added and multiplied in a familiar way: "Write down the remainder and carry the quotient". By way of illustration we give some examples of how one does arithmetic with the 5-adic integers:

$$
\begin{array}{rcr}
\ldots 20134 & \ldots 20134 & \ldots 20134 \\
+\ldots 12203 & \times \ldots 12203 & -\ldots 12203 \\
\hline
\ldots 32342 & \ldots 11012 & \ldots 02431 \\
& \ldots 0000 & \\
& \ldots 323 & \\
& \ldots 23 & \\
& \ldots 4 & \\
& +\ldots & \\
\cline{2-2}
& \ldots 11312 &
\end{array}
$$

We mention in passing that the set $\mathbf{Q}_{p^\infty}$ of p-adic "fractions," which have the form

$$\cdots a_n \cdots a_1 a_0 \cdot a_{-1} \cdots a_{-s} = \sum_{n=-s}^{\infty} a_n p^n, \qquad 0 \le a_n < p,$$

turns out to be a field under the analogous operations; this field is called the *field of p-adic numbers*. The reader will be able to work out for himself the long division procedure for p-adic numbers (it is only for division that p needs to be prime). Thus we have

$$\mathrm{End}\, \mathbf{C}_{p^\infty} \simeq \mathbf{Z}_{p^\infty}, \qquad \mathrm{Aut}\, \mathbf{C}_{p^\infty} \simeq \mathbf{Z}_{p^\infty}^*. \tag{8}$$

As may easily be seen, $\mathbf{Z}_{p^\infty}^*$ consists of those p-adic integers with "residue modulo p" a_0 different from zero.

(III). It is true that

$$\mathrm{Aut}\, \mathbf{S}_n \simeq \mathbf{S}_n, \quad \text{for } n \ne 2, 6. \tag{9}$$

We shall prove this at the end of this section, where we shall also show that 2, 6 are indeed exceptional.

(IV). There is a large body of mathematical literature devoted to describing the automorphisms of the classical matrix groups—see for example [13]. We quote without proof a typical result. Let $n \geq 3$, K a field of characteristic $\neq 2$. Then for each automorphism ϕ in Aut $\mathbf{GL}_n(K)$, either

$$x^\phi = x^\psi \cdot (x^\sigma)^g, \qquad x \in \mathbf{GL}_n(K), \tag{10}$$

or

$$x^\phi = x^\psi \cdot (\hat{x}^\sigma)^g, \qquad x \in \mathbf{GL}_n(K), \tag{11}$$

for some homomorphism $\psi: \mathbf{GL}_n(K) \to K^*$, some $\sigma \in$ Aut K, and some element $g \in \mathbf{GL}_n(K)$. Here the hat denotes the taking of the inverse transpose matrix. For many of the matrix groups, especially over rings, the automorphism groups have been investigated either only a little or not at all.

5.1.2. Exercise. If $|G| > 2$, then Aut $G \neq 1$. (Hint. Consider the cases: 1) G nonabelian; 2) G abelian with an element of order > 2; 3) G satisfying the law $x^2 = e$.)

5.1.3. Exercise. $\mathrm{End}(\mathbf{Z} \oplus \cdots \oplus \mathbf{Z}) \simeq \mathbf{M}_n(\mathbf{Z})$;

$$\mathrm{Aut}(\mathbf{Z} \oplus \cdots \oplus \mathbf{Z}) \simeq \mathbf{GL}_n(\mathbf{Z});$$

$$\mathrm{End}(\mathbf{Z}_m \oplus \cdots \oplus \mathbf{Z}_m) \simeq \mathbf{M}_n(\mathbf{Z}_m);$$

$$\mathrm{Aut}(\mathbf{Z}_m \oplus \cdots \oplus \mathbf{Z}_m) \simeq \mathbf{GL}_n(\mathbf{Z}_m);$$

$$\mathrm{End}(\mathbf{Q} \oplus \cdots \oplus \mathbf{Q}) \simeq \mathbf{M}_n(\mathbf{Q});$$

$$\mathrm{Aut}(\mathbf{Q} \oplus \cdots \oplus \mathbf{Q}) \simeq \mathbf{GL}_n(\mathbf{Q}).$$

(Here n is the number of summands in the left-hand sides.)

5.1.4. Exercise. $\mathrm{End}(\mathbf{C}_{p^\infty} \times \cdots \times \mathbf{C}_{p^\infty}) \simeq \mathbf{M}_n(\mathbf{Z}_{p^\infty})$;

$$\mathrm{Aut}(\mathbf{C}_{p^\infty} \times \cdots \times \mathbf{C}_{p^\infty}) \simeq \mathbf{GL}_n(\mathbf{Z}_{p^\infty});$$

(where n denotes the number of factors in the left-hand sides).

5.1.5. Exercise. From the fact that the orders of an element and its image under an automorphism are the same, list the automorphisms of $\mathbf{S}_3$, and verify that they are all inner.

5.2. Invariant Subgroups

The language of automorphisms allows us to give yet another definition of normal subgroup: A subgroup H of a group G is normal if and only if it admits all the inner automorphisms of the group G; i.e.

$$H^\phi \leq H \quad \text{for all } \phi \in \mathrm{Inn}\, G.$$

If in this condition we replace Inn G by an arbitrary subset Φ of End G, we arrive at a more general concept: A subgroup H of a group G is said to be *invariant* (or sometimes "admissible") with respect to Φ (more briefly Φ-*invariant*), and we write $H \leq_\Phi G$, if

$$H^\phi \leq H \quad \text{for all } \phi \in \Phi.$$

It is obvious that the identity subgroup and the whole group are invariant with respect to arbitrary Φ. If a group contains no other Φ-invariant subgroups than these obvious ones, it is said to be Φ-*simple*. For $\phi \in \Phi$, subsets M, $M\phi$ are termed Φ-*conjugate*. The relation of Φ-conjugacy is utilized only when $\Phi \leq$ Aut G, in which case it is an equivalence relation—the general case is of little use. In the most common situations Φ is either End G, Aut G or Inn G, and then if H is a Φ-invariant subgroup of G, one writes respectively

$$H \leq_e G, \qquad H \leq_a G, \qquad H \leq_i G.$$

Of course more usual than the notation $\leq_i$ is the stylized form $\trianglelefteq$, which we have used from the beginning. It is also more usual that, instead of the term "Φ-invariant" for these Φ, the special terms given in the following table are used. (A better name than "characteristic" might be "automorphically invariant".)

Φ	Φ-conjugate	Φ-invariant	Φ-simple
Inn G	conjugate	normal	simple
Aut G	—	characteristic	characteristically simple
End G	—	fully invariant	—

5.2.1. Exercise. The intersection of a family of Φ-invariant subgroups is again Φ-invariant. The same is true of the subgroup generated by a family of Φ-invariant subgroups.

5.2.2. Exercise. The relations $\leq_e$, $\leq_a$ are transitive (in contrast with the relation $\trianglelefteq$); i.e.

$$A \leq_e B, B \leq_e C \;\Rightarrow\; A \leq_e C;$$

$$A \leq_a B, B \leq_a C \;\Rightarrow\; A \leq_a C.$$

5.2.3. Exercise. A characteristic subgroup of a normal subgroup is normal in the whole group; i.e.

$$A \leq_a B, B \trianglelefteq C \;\Rightarrow\; A \trianglelefteq C.$$

As obvious examples of fully invariant subgroups of a group G, we mention: the successive commutator subgroups; the *nth power* $G^n = \langle x^n \mid x \in G \rangle$; the subgroup $G_n = \langle x \mid x \in G, x^n = 1 \rangle$. The center of a group G is always characteristic since if $ab = ba$, $a, b \in G$, and if $\phi \in \text{Aut } G$, then $a^\phi b^\phi = b^\phi a^\phi$, and as a runs through the group G, its image a^ϕ also runs through the whole group. Note that even in a finite group the center need not be fully invariant—see Example (III) below.

We now give examples of invariant subgroups in various concrete groups.

5.2.4. EXAMPLES. (I). In the groups $\mathbf{Z}, \mathbf{Z}_n$ all subgroups are fully invariant. The group $\mathbf{Q}$ is characteristically simple, since for any pair of nonzero rationals there is an automorphism of $\mathbf{Q}$ mapping one onto the other—see the description of Aut $\mathbf{Q}$ above.

(II). It is clear that

$$\mathbf{C}_p \leq_e \mathbf{C}_{p^2} \leq_e \cdots, \mathbf{C}_{p^n} \leq_e \mathbf{C}_{p^\infty}.$$

(III). Since the commutator subgroup of a group is fully invariant, we have $\mathbf{A}_n \leq_e \mathbf{S}_n$. Let $n \geq 3$. Since $\mathbf{S}_n$ has trivial center, it follows that in the finite group $\mathbf{C}_2 \times \mathbf{S}_n$, $\mathbf{C}_2$ is the centre. It is not fully invariant as it does not admit the endomorphism defined by $(-1)^m x \to (12)^m$, $m = 0, 1, x \in \mathbf{S}_n$.

(IV). Let K be a field. Since the mutual commutator of a pair of Φ-invariant subgroups is clearly Φ-invariant, we have:

$$\mathbf{SL}_n(K) \leq_e \mathbf{GL}_n(K);$$

$$\mathbf{UT}_n^i(K) \leq_e \mathbf{T}_n(K), \qquad i = 1, 2, \ldots.$$

5.2.5. Exercise. The maximal p-subgroup of an abelian group is fully invariant.

5.2.6. Exercise. The Frattini subgroup of an arbitrary group is characteristic.

5.2.7. Exercise. The group $\mathbf{Q} \oplus \cdots \oplus \mathbf{Q}$ is characteristically simple.

Sometimes the following more general situation arises: we are given a group G, and an arbitrary set V with a map $V \to \text{End } G$. Then V is called a *set of operators* acting on G, and for $\Phi \subseteq V$ we extend in the obvious way the above definition of Φ-invariance. For $\Phi = V$ we speak simply of "invariant" subgroups.

5.3. Complete Groups

A group is called *complete* if it has trivial center and all its automorphisms are inner. If a group G is complete then $C(G) = 1$, Out $G = 1$, and by (3), (4)

$$\text{Aut } G \simeq G,$$

so that the study of the automorphism group reduces to the study of the group itself. This property of complete groups is the source of their perhaps imposing name†, given to them at the dawn of group theory. The fact is that complete groups do not play any major role in group theory. (It is analogous to that played by perfect numbers in number theory.) Our reason for introducing the concept is a wish to prove the classical result that most of the symmetric groups are complete.

5.3.1. Theorem (Hölder). *Provided* $n \neq 2$, 6, *the symmetric group* $\mathbf{S}_n$ *is complete.*

PROOF. (i) From §3 we know that for $n \geq 3$ the group $\mathbf{S}_n$ has trivial center. It remains to show that for $n \neq 6$ every automorphism γ of $\mathbf{S}_n$ is inner. Let B_k be the set of all products of k disjoint transpositions from $\mathbf{S}_n$, $1 < k < n/2$. By 2.5.7(III) two permutations from $\mathbf{S}_n$ are conjugate in $\mathbf{S}_n$, if and only if in their disjoint cycle decompositions the number of cycles of each length is the same. Thus in particular the B_k comprise all conjugacy classes of elements of order 2 in $\mathbf{S}_n$, and of course each B_k is mapped onto itself by every inner automorphism of $\mathbf{S}_n$. This suggests the following approach: Prove firstly that γ maps B_1 to itself, and then show that γ is inner.

(ii) To see that γ maps B_1 to itself, note first that since automorphisms preserve the orders of elements, and map conjugacy classes onto conjugacy classes, we have that $B_1^\gamma = B_k$ for some k. We then get the desired equality $B_1^\gamma = B_1$ by brute force: we shall prove that $|B_k| \neq |B_1|$ if $k \neq 1$, $n \neq 6$.

Now $\mathbf{S}_n$ contains $\binom{n}{2}$ transpositions and therefore there are $\prod_{i=0}^{k-1} \binom{n-2i}{2}$ ordered k-tuples of disjoint transpositions. Since the order of the factors in a product of disjoint transpositions is immaterial, we have that

$$|B_k| = \frac{1}{k} \prod_{i=0}^{k-1} \binom{n-2i}{2} = \frac{1}{k!2^k} n(n-1) \cdots (n-2k+1).$$

Hence the equation $|B_1| = |B_k|$ becomes:

$$(n-2)(n-3) \cdots (n-2k+1) = k!2^{k-1}. \tag{12}$$

We prove that for $n \neq 6$, $k \neq 1$ this is impossible. Since the right-hand side of (12) is positive we must have $n \geq 2k$. (This was in any case the situation pertaining to the theorem.) This implies the following inequality involving the left–hand side of (12):

$$(n-2)(n-3) \cdots (n-2k+1) \geq (2k-2)!.$$

It is easily proved by induction that

$$(2k-2)! > k!2^{k-1}, \quad \text{for } k \geq 4.$$

---

† The Russian name would be more appropriately translated as "perfect." This is reserved in English however for groups equal to their derived group.

It remains to consider the cases $k = 2, 3$. For $k = 2$ it is easy to verify that the equation (12) does not hold for any n. Let $k = 3$. Since $n \geq 2k$ and $n \neq 6$, we have $n > 6$, and

$$\text{l.h.s. } (12) \geq 5.4.3.2 > 3!2^2 = \text{r.h.s. } (12).$$

Thus (12) does not hold for $n \neq 6$, $k \neq 1$.

(iii) To prove that γ is inner we shall define, by induction on r, inner automorphisms $\gamma_2, \ldots, \gamma_r$ with the property that $\gamma\gamma_2^{-1} \cdots \gamma_r^{-1}$ fixes each of the transpositions $(12), \ldots, (1r)$. Then when $r = n$ we get an automorphism $\gamma\gamma_2^{-1} \cdots \gamma_n^{-1}$ fixing every transposition $(ij) = (1i)(1j)(1i)$. Since $\mathbf{S}_n$ is generated by its transpositions we shall then have that $\gamma\gamma_2^{-1} \cdots \gamma_n^{-1} = 1$, so that γ is inner.

First take $r = 2$. By Part (i) above, $(12)\gamma = (ij)$ for some i, j. Define γ_2 to be any inner automorphism with the property that $(12)^{\gamma_2} = (ij)$. Then $\gamma\gamma_2^{-1}$ fixes the transposition (12).

For the inductive step, suppose $\gamma' = \gamma\gamma_2^{-1} \cdots \gamma_r^{-1}$ fixes each of the transpositions $(12), \ldots, (1r)$, and that $(1, r+1)^{\gamma'} = (ij)$. The intersection $\{1, 2\} \cap \{i, j\}$ is not empty, since if it were, then from the equation

$$((12)(1, r+1))^{\gamma'} = (12)(ij)$$

we should get that an element of order 3 is mapped by γ' to an element of order 2. We may therefore assume that $i = 1$ or $i = 2$, whence it follows that $j > r$ since γ' fixes $(12), \ldots, (1r)$ and

$$(ij) = \begin{cases} (1j) & \text{for } i = 1; \\ (12)(1j)(12) & \text{for } i = 2. \end{cases}$$

Consider first the case $r \geq 3$. The intersection $\{1, 3\} \cap \{i, j\}$ is also nonempty, whence $i = 1$. Define γ_{r+1} to be conjugation by the permutation $(r+1, j)$; note that $\gamma_{r+1} = 1$ if $j = r+1$. Then γ_{r+1} acts on $(12), \ldots, (1, r+1)$ in the same way as γ', so that $\gamma'\gamma_{r+1}^{-1} = \gamma\gamma_2^{-1} \cdots \gamma_{r+1}^{-1}$ fixes the transpositions $(12), \ldots, (1, r+1)$. It remains to look at the case $r = 2$. If the intersection $\{1, 2\} \cap \{i, j\}$ contains 1 then the preceding argument still applies. In the contrary situation we may assume that $(ij) = (23)$ or (24), whence we have that $(12)^{\gamma'} = (12)$, $(13)^{\gamma'} = (23)$ or (24). Denote by γ_3 conjugation by (12) or $(12)(34)$ respectively. Obviously γ_3 acts on (12) and (13) in the same way as γ', so that $\gamma'\gamma_3^{-1} = \gamma\gamma_2^{-1}\gamma_3^{-1}$ fixes the transpositions (12), (13). This completes the proof of the theorem.

The groups $\mathbf{S}_2$, $\mathbf{S}_6$ are not complete since the first is abelian and the second has an outer automorphism of order 2. The definition of this automorphism may be found in for example a note by D. W. Miller [On a theorem of Hölder, Amer. Math. Monthly **65**, 4 (1958), 252–254].

5.3.2. Exercise. A complete normal subgroup is always a direct factor of its supergroup. (A leading question is: What must the other factor be?)

§6. Extensions by Means of Automorphisms

We describe two group-theoretically important constructions which use automorphisms.

6.1. The Holomorph

This construction arises in connexion with the following question: Is it possible to embed an arbitrary group G in some group G^* with the property that every automorphism of G is the restriction of some inner automorphism of G^*? Write briefly $\Phi = \operatorname{Aut} G$. Then for G^* one may take the set of ordered pairs ϕg, $\phi \in \Phi$, $g \in G$, with multiplication defined by the rule:

$$\phi g \cdot \phi' g' = \phi \phi' g^{\phi'} g'. \tag{1}$$

(We are writing pairs without their customary comma and brackets.) The group axioms are straightforward to verify. It is also straightforward that the maps

$$\Phi \to G^*, \qquad G \to G^*, \tag{2}$$

defined by $\phi \to \phi 1$, $g \to 1g$, are embeddings (i.e. monomorphisms). We identify Φ and G with their images in G^* under these monomorphisms. From the rule for multiplication (1) it is immediate that

$$\phi^{-1} g \phi = g^{\phi} \quad \text{for } \phi \in \Phi, \qquad g \in G. \tag{3}$$

It is then clear that

$$G^* = \Phi G, \qquad G \lhd G^*, \qquad \Phi \cap G = 1, \tag{4}$$

and by (3) every automorphism $\phi \in \Phi$ is the restriction of some inner automorphism of the group G^*. Hence our problem is solved. The group ΦG is called the *holomorph* of the group G, and is denoted by $\operatorname{Hol} G$.

If instead of taking $\Phi = \operatorname{Aut} G$ we take $\Phi \leq \operatorname{Aut} G$, then the group ΦG will as before have properties (3), (4). In this more general situation ΦG is called the *extension of the group G by means of the automorphisms in Φ*. By (4) this accords with the general definition of extension given in §4.4.

The situation may be generalized yet further by taking Φ to be any group which comes equipped with a homomorphism $\Phi \to \operatorname{Aut} G$. By defining conjugation of G by an element of Φ to have the same effect as the corresponding automorphism, and using the same multiplication (1), we turn ΦG into a group having properties (3), (4). In this case ΦG is called the *extension of the group G by means of the group of operators Φ*.

We now look at the holomorphs of some concrete groups.

6.1.1. EXAMPLES. (I). Let K be any of the rings $\mathbf{Z}$, $\mathbf{Z}_n$, $\mathbf{Q}$. By 5.1.1(I) the automorphisms of the additive group K are just the multiplications by elements of K^*, so that

$$\text{Hol } K \simeq \left\{ \begin{pmatrix} 1 & \beta \\ 0 & \alpha \end{pmatrix} \middle| \alpha \in K^*, \beta \in K \right\}.$$

(II). The elements of Hol $\mathbf{C}_{p^\infty}$ can be written in a similar form. Call a partial sequence $(s_1, s_2, \ldots)$ a p-*thread* if the first few places are empty while in the remainder there are elements $s_n \in \mathbf{Z}_{p^n}$, satisfying $ps_n = s_{n+1}$ (suitably interpreted), and if also the empty places cannot be filled so as to satisfy these conditions. The sum of two p-threads is defined as follows: Add them componentwise where they both *have* components and then fill in the blanks wherever possible in conformance with the conditions of the definition of p-thread. It is easily verified that with this operation the set of all p-threads is a group (the "direct limit" of the $\mathbf{Z}_{p^n}$). It is in fact isomorphic to $\mathbf{C}_{p^\infty}$. To see this, for each $n = 1, 2, \ldots$, choose a primitive complex p^nth root of unity ε_n such that $\varepsilon_{n+1}^p = \varepsilon_n$. Each complex number x in $\mathbf{C}_{p^\infty}$ lies in some smallest $\mathbf{C}_{p^n}$, and then in all succeeding $\mathbf{C}_{p^m}$, $m \geq n$. Write $x = \varepsilon_m^{s_m}$, $s_m \in \mathbf{Z}_{p^m}$, for all $m \geq n$. Since $\varepsilon_{m+1}^{s_{m+1}} = \varepsilon_m^{s_m} = \varepsilon_{m+1}^{ps_m}$, it follows that the partial sequence of the exponents s_m is a p-thread. It is easy to see that the map $x \to (s_1, s_2, \ldots)$ is an isomorphism from the group $\mathbf{C}_{p^\infty}$ onto the group of p-threads, and moreover that the automorphisms in Aut $\mathbf{C}_{p^\infty}$ when represented as in 5.1.1(II) as p-adic integers, act on the p-threads by componentwise multiplication. Thus

$$\text{Hol } \mathbf{C}_{p^\infty} \simeq \left\{ \begin{pmatrix} 1 & \beta \\ 0 & \alpha \end{pmatrix} \middle| \alpha \in \mathbf{Z}_p^*, \beta \text{ a } p\text{-thread} \right\}.$$

This differs from the preceding example in that the matrix entries α, β come from two different sets with no ring evident containing them both.

(III). We know that for $n \neq 2, 6$, the group $\mathbf{S}_n$ is complete so that certainly Aut $\mathbf{S}_n \simeq \mathbf{S}_n$. Since $\mathbf{S}_n$ is normal in its holomorph it is, by Exercise 5.3.2, a direct factor of its holomorph. Hence

$$\text{Hol } \mathbf{S}_n \simeq \mathbf{S}_n \times \mathbf{S}_n \quad \text{for } n \neq 2, 6.$$

It is clear that in fact for any complete group G this is true: Hol $G \simeq G \times G$.

(IV). For the holomorphs of the groups $\mathbf{GL}$, $\mathbf{SL}$, $\mathbf{T}$ etc. one cannot hope perhaps for a description simpler than the definition, since the automorphism groups in these cases have rather complex structures.

6.1.2. Exercise. Let K be any of the rings $\mathbf{Z}$, $\mathbf{Z}_m$, $\mathbf{Q}$. Establish the isomorphism (n summands in both sums):

$$\text{Hol}(K \oplus \cdots \oplus K) \simeq \left\{ \begin{pmatrix} 1 & \beta \\ 0 & \alpha \end{pmatrix} \middle| \alpha \in \mathbf{GL}_n(K), \beta \in K \oplus \cdots \oplus K \right\}.$$

6.1.3. Exercise. Find all the conjugacy classes of Hol $\mathbf{Z}$.

6.1.4. Exercise. Find in Hol $\mathbf{Z}$ a chain

$$\text{Hol } \mathbf{Z} = H_0 \geq H_1 \geq \cdots$$

of normal subgroups H_i such that $\bigcap H_i = 1$ and each factor H_i/H_{i+1} is contained in the center of the factor group H_0/H_{i+1}, $i = 1, 2, \ldots$.

6.1.5. Exercise. Is Hol $\mathbf{Z}$ a complete group?

6.2. Wreath Products

Let A, B be groups. We denote by $A^{[B]}$, $A^{(B)}$ the Cartesian and direct products respectively of isomorphic copies of A indexed by the elements of B. Thus $A^{[B]}$ is the group of all functions $B \to A$ with the usual multiplication, while $A^{(B)}$ is the subgroup consisting of all such functions with finite support. For each $f \in A^{[B]}$, $b \in B$ we define a function f^b by

$$f^b(x) = f(bx), \qquad x \in B. \tag{5}$$

It can be verified immediately that the map

$$\hat{b} : A^{[B]} \to A^{[B]}, \tag{6}$$

defined by $f \to f^b$, is an automorphism mapping $A^{[B]}$ onto itself, and that the maps

$$B \to \operatorname{Aut}(A^{[B]}), \qquad B \to \operatorname{Aut}(A^{(B)}), \tag{7}$$

sending each $b \in B$ to the automorphism $\hat{b}$ and to the restriction of $\hat{b}$ to $A^{(B)}$ respectively, are monomorphisms. The extensions of the groups $A^{[B]}$, $A^{(B)}$ by means of the groups of operators (7) are called respectively the (*unrestricted*) *wreath product* and *restricted wreath product of the group A by the group B*, and are denoted by A Wr B, A wr B. Thus the unrestricted wreath product A Wr B is the set product $B \times A^{[B]}$ with multiplication given by

$$bf \cdot b'f' = bb'f^{b'}f', \quad \text{where } f^{b'}(x) = f(b'x), \tag{8}$$

while the restricted wreath product is its subgroup $B \cdot A^{(B)}$. Notice that in the construction of the wreath products of A by B, the groups A and B play different roles: A is *passive* and B is *active*. In English the names "bottom" and "top" groups (for A, B respectively) are more usual: however the names we adopt here are more descriptive and so more suitable.

The subgroups $A^{[B]}$, $A^{(B)}$ are called the *base groups* of the corresponding wreath products. It is obvious that A Wr B and A wr B coincide if and only if either A is trivial or B finite. The subgroup

$$\operatorname{Diag}(A^{[B]}) = \{f | f \in A^{[B]}, f(x) = \text{const. for } x \in B\};$$

i.e. the diagonal of the base group $A^{[B]}$, is called also the *diagonal subgroup of the wreath product A Wr B*. Finally, for each $a \in A$ we define a function $\bar{a} \in A^{[B]}$, by

$$\bar{a}(x) = \begin{cases} a & \text{for } x = e, \\ e & \text{for } x \neq e. \end{cases}$$

It is immediate that the map $A \to A^{[B]}$ defined by $a \to \bar{a}$ is a monomorphism. The image $\bar{A}$ of A under this map is called the *first coordinate subgroup*, and for $b \in B$ the conjugate subgroup $\bar{A}^b$ is called the bth *coordinate subgroup* of either wreath product. Thus the passive group A participates in the wreath product through its copies $\bar{A}^b$, while the active group B is contained in the wreath product, and by acting by conjugation on the copies of A—the coordinate subgroups—permutes or "wreathes" them. Obviously

$$A^{(B)} = \prod_{b \in B}^{\times} \bar{A}^b.$$

6.2.1. Exercise. If $A_1 \leq A$, $B_1 \leq B$, then the wreath product $A_1 \text{ wr } B_1$ is isomorphic to the subgroup of $A \text{ wr } B$ generated by the image of A_1 in the first coordinate subgroup $\bar{A}$ under the canonical isomorphism $A \to \bar{A}$, and the subgroup B_1.

6.2.2. Exercise. A nontrivial normal subgroup of a restricted wreath product with nontrivial passive group has nontrivial intersection with the base group.

6.2.3. Exercise. Let A be a nontrivial group. Then

$$C(A \text{ Wr } B) = \text{Diag}(C(A)^{[B]});$$

$$C(A \text{ wr } B) = 1, \quad \text{if } B \text{ is infinite.}$$

6.2.4. Exercise. The commutator subgroup of the wreath product $A \text{ wr } B$ is the product $B' \cdot H$, where

$$H = \left\{ f \mid F \in A^{(B)}, \prod_{b \in B} f(b) \equiv e \bmod A' \right\}.$$

(As usual A', B' denote $[A, A]$, $[B, B]$.)

6.2.5. Exercise. Every element of the commutator subgroup of $\mathbf{Z} \text{ wr } \mathbf{Z}$ is a commutator.

6.2.6. Exercise. The wreath product $\mathbf{Z} \text{ wr } \mathbf{Z}$ is isomorphic to the subgroup of $\mathbf{GL}_n(\mathbf{R})$ generated by the matrices

$$\begin{pmatrix} 1 & 1 \\ 0 & 1 \end{pmatrix}, \quad \begin{pmatrix} \zeta & 0 \\ 0 & 1 \end{pmatrix},$$

where ζ is any fixed transcendental real.

6.2.7. Exercise. The operations Wr, wr are not associative on the class of all groups (in fact not even on the class of finite groups).

An important link between wreath products and arbitrary extensions is given by the following result, dating back to Frobenius.

6.2.8. Theorem (Kalužnin–Krasner). *Every extension of a group A by a group B can be embedded in the unrestricted wreath product $A \operatorname{Wr} B$.*

PROOF. Let $A \trianglelefteq G$, $G/A = B$, $\tau: B \to G$, a function (a "coset representative function") choosing for each coset $b \in B$, an element in that coset (τ for "transversal"). Write briefly $W = A \operatorname{Wr} B$. Define a map $\phi_\tau: G \to W$, by $g^{\phi_\tau} = \hat{g} f_g$, $g \in G$, where $\hat{g}$ is the coset Ag, and f_g is the element of the base group of W defined by

$$f_g(b) = ((\hat{g}b)\tau)^{-1} g(b\tau), \qquad b \in B.$$

A direct computation verifies that ϕ_τ is a monomorphism. This proves the theorem.

6.2.9. Exercise. With the same notation,

$$G^{\phi_\tau} \cdot A^{[B]} = W, \qquad G^{\phi_\tau} \cap A^{[B]} = A^{\phi_\tau}.$$

6.2.10. Exercise. If τ_1 is another coset representative function then the subgroups G^{ϕ_τ}, $G^{\phi_{\tau_1}}$ are conjugate in W.

We remark that $A \operatorname{Wr} B$ is sometimes referred to as the complete or Cartesian wreath product, while $A \operatorname{wr} B$ is also called the discrete or direct wreath product. We shall generalize slightly the concept of wreath product as defined here, when we consider (in §13) monomial representations—it was in this guise that the wreath product first made its appearance on the historical scene. There we shall describe an embedding of an arbitrary group G with a given subgroup H of prescribed index m, into a monomial group of matrices of degree m over H—an early version of the theorem proved above.

Finally note that the holomorph and the wreath product (and the direct product of two groups) are rather special extensions of one group by another in the following sense: An extension G of a group A by a group B is said to *split*, or be a *splitting extension*, if G contains subgroups A_1, B_1 isomorphic to A, B respectively such that $A_1 \trianglelefteq G$, $G = A_1 B_1$, and $A_1 \cap B_1 = 1$. Clearly we then have that $G/A_1 \cong B$. Alternatively one sometimes says that A_1 is *complemented* in G (by B_1), or that G is a *semidirect product* of A by B.

6.2.11. Exercise. Every extension of $\mathbf{Z}_p$ by $\mathbf{Z}_q$, where p, q are distinct primes, is splitting. The group $\mathbf{Z}_{p^2}$ is an extension of $\mathbf{Z}_p$ by itself which is not splitting.

3 Abelian Groups

For abelian groups it is both more convenient and more usual to use additive notation. We shall follow this convention in the present chapter.

§7. Free Abelian Groups. Rank

7.1. Free Abelian Groups

Let $\mathfrak{L}$ be a class of groups. A group $F = \langle x_i \mid i \in I \rangle$ belonging to $\mathfrak{L}$ is said to be *free in the class $\mathfrak{L}$, freely generated by the set* $\{x_i \mid i \in I\}$, if for every group $G \in \mathfrak{L}$ with generating set $\{a_i \mid i \in I\}$ the map $x_i \to a_i$ extends to a homomorphism $F \to G$. The cardinal of the index set I is called the *rank* of the free group F. The set $\{x_i \mid i \in I\}$ is often called a *basis* for F. It may be shown easily that not every class of groups contains free groups. However free groups do exist in the class of all abelian groups, and can be described very simply. Before giving this description we need the following preparatory lemma.

7.1.1. Lemma. *Suppose a factor group G/N of an abelian group G decomposes as a direct sum of infinite cyclic groups:*

$$G/N = \bigoplus_{i \in I} (A_i/N), \qquad A_i = \langle a_i, N \rangle.$$

Then G is the direct sum of the subgroups N and $A = \langle a_i \mid i \in I \rangle$.

Proof. First observe that $G = \langle N, A \rangle$. Next suppose that $A \cap N \neq 0$, and let a be a nonzero element in $A \cap N$. Then a can be written in the form

$$a = \sum n_k a_{i_k}, \qquad n_k \in \mathbf{Z}.$$

Hence we get an equation in G/N:

$$N = \sum (n_k a_{i_k} + N),$$

whence by the definition of direct sum it follows that $n_k a_{i_k} + N = N$. Since A_{i_k}/N is infinite cyclic we must therefore have that $n_k = 0$; but this means that $a = 0$, a contradiction.

7.1.2. Theorem. *The free groups in the class of abelian groups are precisely the direct sums of infinite cyclic groups.*

PROOF. Let $G = \bigoplus_{i \in I} \langle x_i \rangle$ be a direct sum of infinite cyclic groups $\langle x_i \rangle$, and let A be any group with generators a_i, $i \in I$. It is easy to see that the map $\sum n_k x_{i_k} \to \sum n_k a_{i_k}$, which extends the map $x_i \to a_i$ of the set $\{x_i \mid i \in I\}$ onto the set $\{a_i \mid i \in I\}$, is a homomorphism $G \to A$. This means that G is a free abelian group, i.e. a free group in the class of abelian groups. Its rank is the number of infinite cyclic direct factors.

Now let F be a free abelian group and $\{x_i \mid i \in I\}$ a set of free generators for it. By the definition of free group there exists a homomorphism τ from F onto the direct sum $A = \bigoplus_{i \in I} \langle a_i \rangle$ say, of infinite cyclic groups $\langle a_i \rangle$, which extends the map $x_i \to a_i$. By one of the homomorphism theorems $A \simeq F/N$ where $N = \mathrm{Ker}\ \tau$, so that the factor group F/N decomposes as the direct sum of the infinite cyclic groups $\langle x_i + N \rangle$, $i \in I$. Then by 7.1.1 we have $F = N \oplus B$ where $B = \langle x_i \mid i \in I \rangle$. However the subgroup B is generated by the same set $\{x_i \mid i \in I\}$ as F; thus the kernel N of the homomorphism τ is zero, i.e. τ is an isomorphism (being obviously onto). Hence our free abelian group F is isomorphic to A, the direct sum of infinite cyclic groups, and the theorem is proved.

From the proof of Theorem 7.1.2 we see that the rank of a free abelian group G is independent of the choice of basis, and coincides with the number of summands in the decomposition of G as a direct sum of cyclic subgroups. (Prove that this number is an invariant of G.)

It turns out that the subgroups of a free abelian group are again free abelian. For the proof of this fact we shall make use of the concept of *ascending series*

$$0 = N_0 < N_1 < \cdots < N_\alpha < \cdots < N_\gamma = G \qquad (1)$$

of an abelian group G, i.e. a chain of subgroups well-ordered by inclusion and indexed by the corresponding ordinal numbers, with the additional stipulation that for each limit ordinal α the subgroup N_α is the union of all subgroups N_β, $\beta < \alpha$. The *factors* of the series (1) are the quotients $N_{\alpha+1}/N_\alpha$.

If $A \leq G$ then in the series

$$0 = A_0 \leq A_1 \leq \cdots \leq A_\alpha \leq \cdots \leq A_\gamma = A, \qquad (2)$$

where $A_\alpha = A \cap N_\alpha$, there may be repetitions. After omitting duplications and renumbering the remaining (distinct) members of (2), we obtain an ascending series

$$0 = \hat{A}_0 < \hat{A}_1 < \cdots < \hat{A}_\rho < \cdots < \hat{A}_\tau = A \qquad (3)$$

of subgroups of A, which we shall say has been obtained by intersecting the series (1) with the subgroup A.

To prove the subgroup theorem we require the following condition equivalent to freeness of an abelian group.

7.1.3. Theorem. *An abelian group G is free if and only if it possesses an ascending series all of whose factors are infinite cyclic.*

PROOF. Suppose G has an ascending series (1) with infinite cyclic factors. For each $\alpha < \gamma$ choose in the set difference $N_{\alpha+1} \backslash N_\alpha$ an element $a_{\alpha+1}$ such that $N_{\alpha+1} = \langle a_{\alpha+1} + N_\alpha \rangle$; we shall show that G is the direct sum $\bigoplus_{\alpha < \gamma} \langle a_{\alpha+1} \rangle$ of its infinite cyclic subgroups $\langle a_{\alpha+1} \rangle$. The proof will be by induction on γ, the length of the series (1). For $\gamma = 1$ the assertion is obvious; assume inductively that it holds for all $\alpha < \gamma$.

Let g be any nonzero element of G, and suppose $g \in N_\beta$, $g \notin \bigcup_{\sigma < \beta} N_\sigma = N_{\beta-1}$ (in other words β is the smallest ordinal such that $g \in N_\beta$, and so cannot be a limit ordinal). Since $N_\beta = \langle a_\beta, N_{\beta-1} \rangle$ and $N_\beta / N_{\beta-1}$ is infinite cyclic, g can be written uniquely in the form $g = g_1 + n a_\beta$, $g_1 \in N_{\beta-1}$. By the inductive hypothesis, since $\beta - 1 < \gamma$, the element g has a unique expression as a linear combination of the a_{β_i}:

$$g_1 = n_1 a_{\beta_1} + \cdots + n_s a_{\beta_s}, \qquad \beta_i < \beta,$$

which implies the uniqueness of the expression

$$g = n_1 a_{\beta_1} + \cdots + n_s a_{\beta_s} + n a_\beta.$$

This shows that G is a direct sum of infinite cyclic groups, which by Theorem 7.1.2 is equivalent to freeness.

For the converse let G be a free abelian group and let $G = \bigoplus_{\alpha < \gamma} \langle g_\alpha \rangle$ be a decomposition of G as a direct sum of infinite cyclic groups. Setting $N_0 = 0$, $N_{\alpha+1} = \langle g_\alpha, N_\alpha \rangle$ and $N_\alpha = \bigcup_{\beta < \alpha} N_\beta$ for limit ordinals α, we obtain an ascending series with infinite cyclic factors. This shows the necessity of the condition and completes the proof of the theorem.

In §4.4 we showed that the factors of a subnormal series of a subgroup $A \le G$, obtained by intersecting A with the terms of a subnormal series of G, are isomorphic to subgroups of the factors of the series for G. The corresponding assertion for the ascending series (1) and (3) is established in exactly the same way. From this remark and Theorem 7.1.3 the subgroup theorem for free abelian groups follows immediately:

7.1.4. Theorem. *Every subgroup (including the zero subgroup) of a free abelian group is again free abelian.*

7.1.5. Exercise. Let n be a positive integer and let $\mathfrak{A}_n$ be the class of all abelian groups satisfying the law $nx = 0$. Prove that the free groups of the class $\mathfrak{A}_n$ are just the direct sums of isomorphic copies of $\mathbf{Z}_n$. For which of the classes $\mathfrak{A}_n$ is it true that subgroups of free groups are always free?

7.2. Rank of an Abelian Group

In the preceding section we defined the rank of a free abelian group to be the cardinal of a set of free generators. As it stands this definition makes sense only for *free* abelian groups; however we shall now give another definition of rank applicable to arbitrary abelian groups. As will be apparent this new rank coincides with the old rank in the case of free abelian groups, and is analogous to the dimension of a vector space.

A finite family of elements $g_1, \ldots, g_k$ of an abelian group C is said to be *linearly dependent* (over $\mathbf{Z}$) if there exist integers $n_1, \ldots, n_k$, not all zero, such that $\sum n_i g_i = 0$. An arbitrary family of elements of the group G is *linearly dependent* if some finite subfamily is linearly dependent. It is easy to show that in an abelian group G containing at least one element of infinite order there is always a maximal linearly independent subset, and that all such subsets have the same cardinal. This cardinal, the size of any maximal linearly independent subset of G, is what we shall call the *rank* of the group G.

Periodic abelian groups obviously do not contain linearly independent subsets (except for the empty set), so that it is natural to define their ranks to be zero.

7.2.1. Theorem. *A nonzero torsion-free abelian group has rank* 1 *if and only if it is isomorphic to a subgroup of the additive group* $\mathbf{Q}$ *of rationals.*

PROOF. Let $G \le \mathbf{Q}$, $G \ne 0$, and let g_1, g_2 be any nonzero elements of G. Then there exist integers $n_1, n_2 \ne 0$ such that $n_1 g_1 = n_2 g_2$. Thus any two elements of G form a linearly dependent family, so that G has rank 1.

Now suppose that we are given a torsion-free group G of rank 1. Fix arbitrarily on some nonzero element $g_0 \in G$. Then for each element $g \in G$ there are integers m, n (where $n \ne 0$ and if $g \ne 0$ then also $m \ne 0$), such that $ng = mg_0$. Choosing such integers m, n for each g, we get a map $\phi: g \to m/n$, from the group G to the group $\mathbf{Q}$.

We show first that in fact m/n is uniquely determined by g. Suppose $n_1 g = m_1 g_0$. Multiplying the equations $ng = mg_0$ and $n_1 g = m_1 g_0$ by m_1 and m respectively, and subtracting we get $(nm_1 - n_1 m)g = 0$. If $g \ne 0$, then $nm_1 - n_1 m = 0$, or $m/n = m_1/n_1$, as desired. If $g = 0$ then we must have that m, m_1 are zero, whence $m/n = m_1/n_1 = 0$.

To show that ϕ is one-to-one, suppose that $kg_1 = lg_0$ where $l/k = m/n$. Then from the equations $ng = mg_0$ and $kg_1 = lg_0$ we get $mk(g - g_1) = 0$, and then the torsion-freeness gives $g = g_1$.

Finally, ϕ is homomorphic: From the equations $ng = mg_0$, $sg' = tg_0$, $g, g' \in G$, it follows that $ns(g + g') = (sm + nt)g_0$, whence $(g + g')\phi = g\phi + g'\phi$. This completes the proof.

Prompted by the well-known theorem about the decomposition of a vector space as a direct sum of one-dimensional subspaces, we might ask: Is every torsion-free abelian group a direct sum of rank 1 abelian groups? The answer is in general negative. An example of an indecomposable torsion-free abelian group of rank 2 was first constructed by L. S. Pontrjagin (see [34], Example 15). There now exists a whole series of papers on the problem of decomposability of torsion-free abelian groups, in which in particular other such examples are given.

7.2.2. Exercise. Let G be an abelian group and let $A \le G$. Then the rank of G is the sum of the ranks of A and G/A.

§8. Finitely Generated Abelian Groups

In this section we first prove a sharpened version of the subgroup theorem for free abelian groups of finite rank, and then from this sharpened result deduce the fundamental theorem about finitely generated abelian groups.

8.1.1. Theorem. *Let F_n be a free abelian group of finite rank n and let A be a nonzero subgroup. Then A is free, and the groups A and F_n possess bases $\{a_1, \ldots, a_k\}$ and $\{f_1, \ldots, f_n\}$ respectively, satisfying the following conditions: $k \le n$; $a_i = m_i f_i$, $1 \le i \le k$; and m_i divides m_{i+1}, $1 \le i \le k - 1$.*

PROOF. We shall use induction on the rank n of the free abelian group F_n. For $n = 1$ the group F_n is cyclic and in this case the statement of the theorem is obviously true. Suppose that $n > 1$, and as inductive hypothesis that the statement is true for free abelian groups of rank $n - 1$.

Given an (ordered) basis $\{x_1, \ldots, x_n\}$ of F_n and a nonzero element $a \in A$, there is an n-tuple $(t_1, \ldots, t_n)$ of integers uniquely determined by the equation

$$a = t_1 x_1 + \cdots + t_n x_n.$$

From all such n-tuples choose one with smallest positive t_1, say $(m_1, s_2, \ldots, s_n)$, $(t_1 = m_1)$, and let

$$\{f_1', b_2, \ldots, b_n\} \tag{1}$$

be a corresponding (ordered) basis, and $0 \neq a_1$ be the element of A such that

$$a_1 = m_1 f_1' + s_2 b_2 + \cdots + s_n b_n.$$

It turns out that under these conditions m_1 divides all the coefficients s_i. For, writing $s_i = q_i m_1 + r_i$, $0 \leq r_i < m_1$, and writing a_1 in terms of the basis

$$\{f_1 = f_1' + q_2 b_2 + \cdots + q_n b_n, b_2, \ldots, b_n\}, \tag{2}$$

we have $a_1 = m_1 f_1 + r_2 b_2 + \cdots + r_n b_n$, so that by the choice of m_1 all the r_i must be zero. Thus $a_1 = m_1 f_1$.

Write $B = A \cap F_{n-1}$, where $F_{n-1} = \langle b_2, \ldots, b_n \rangle$. We shall prove that A is the direct sum of its subgroups $\langle a_1 \rangle$ and B. Since $\langle a_1 \rangle \cap B = 0$ it suffices to show that $A = \langle a_1, B \rangle$.

If $a = mf_1 + b \in A$, where $b \in F_{n-1}$, and $m = qm_1 + r$, $0 \leq r < m_1$, then the element $a - qa_1$ belongs to A and in its expression in terms of the basis (2), the coefficient of f_1 is $r < m_1$, so that again $r = 0$. It follows that $b = a - mf_1 = a - qa_1 \in A$, whence $b \in B$. Since a was an arbitrary element of A we get that

$$A = \langle a_1 \rangle \oplus B.$$

By the inductive hypothesis the subgroup B ($\leq F_{n-1}$) and the group F_{n-1} possess bases $\{a_2, \ldots, a_k\}$ and $\{f_2, \ldots, f_n\}$ respectively, satisfying: $k \leq n$; $a_i = m_i f_i$, $2 \leq i \leq k$; $m_i \mid m_{i+1}$, $2 \leq i < k$. Obviously the sets $\{a_1, a_2, \ldots, a_k\}$ and $\{f_1, f_2, \ldots, f_n\}$ are bases for A and F_n respectively. To show that these bases satisfy the conditions of the theorem it only remains to prove that m_1 divides m_2.

Let $m_2 = \hat{q} m_1 + \hat{r}$, $0 \leq \hat{r} < m_1$. Expressing the element $a_2 - a_1 \in A$ in terms of the new basis $\{\hat{q} f_2 - f_1, f_2, \ldots, f_n\}$, we have

$$a_2 - a_1 = m_1(\hat{q} f_2 - f_1) + \hat{r} f_2.$$

Since the coefficient of f_2 is $\hat{r} < m_1$, we conclude as before that $\hat{r} = 0$, so that m_1 divides m_2 as required. This completes the proof of the theorem.

8.1.2. Theorem. *Every finitely generated abelian group is a direct sum of cyclic subgroups.*

PROOF. Let G be a finitely generated abelian group generated by n elements; then G is isomorphic to a quotient F_n/A of the free group F_n of rank n. By Theorem 8.1.1 the groups F_n and A possess bases $f_1, \ldots, f_n$ and $a_1, \ldots, a_k$ with the property that $a_i = m_i f_i$, $1 \leq i \leq k$. Since $G \simeq F_n/A$ it suffices for the proof of the theorem to show that F_n/A is the direct sum of its cyclic subgroups $\langle f_i + A \rangle$.

In the first place it is clear that F_n/A is generated by the subgroups $\langle f_i + A \rangle$. Next suppose that the zero of the factor group F_n/A can be written

$$A = l_1 f_1 + \cdots + l_n f_n + A.$$

From this we get that $l_1 f_1 + \cdots + l_n f_n = a \in A$. Writing the element a in terms of the above basis for A and utilizing the equations $a_i = m_i f_i$, we arrive at the following equalities:

$$l_1 f_1 + \cdots + l_n f_n = s_1 a_1 + \cdots + s_k a_k = s_1 m_1 f_1 + \cdots + s_k m_k f_k.$$

By the uniqueness of the expression of an element in terms of the free generators f_i, we therefore have: $l_i = s_i m_i$, $1 \le i \le k$; $l_j = 0$, $k < j \le n$. However this means that all of the elements $l_i f_i$ belong to A, i.e. $l_i f_i + A = A$. This gives us the uniqueness of the representation of 0 as a sum of elements of the subgroups $\langle f_i + A \rangle$.

8.1.3. Exercise. Let F be a free abelian group of finite rank n with a basis

$$\{f_1, \ldots, f_n\} \tag{3}$$

We define an *elementary (Nielsen) transformation* of the basis (3) to be a map (or "change of basis") of one of the following kind:

   (i) For a pair i, j with $i \ne j$, $f_i \to f_j$, $f_j \to f_i$, $f_k \to f_k$, $k \ne i, j$;
  (ii) For a pair i, j with $i \ne j$, $f_i \to f_i + f_j$; $f_k \to f_k$, $k \ne i$;
 (iii) For some i, $f_i \to -f_i$; $f_k \to f_k$, $k \ne i$.

Show that every ordered basis for F can be obtained from the basis (3) by a succession of elementary transformations.

8.1.4. Exercise. The collection of all finite index subgroups of a finitely generated abelian group has zero intersection.

8.1.5. Exercise. The set of all elements of finite order of a finitely generated abelian group constitutes a finite subgroup.

8.1.6. Exercise. Show by means of an example that Theorem 8.1.1 does not generalize to free abelian groups of infinite rank.

8.1.7. Exercise. Every subgroup of a finitely generated abelian group is finitely generated.

§9. Divisible Abelian Groups

A group G is said to be *divisible* (or sometimes "radicable") if for every positive integer n and every element $g \in G$, the equation $nx = g$ ($x^n = g$ in multiplicative notation) has at least one solution in G.

9.1.1. EXAMPLES. (I). It is obvious that the additive group $\mathbf{Q}$ of rationals is divisible.

(II). We shall prove the divisibility of the quasicyclic group C_{p^∞}. As we have seen C_{p^∞} is isomorphic to the union of an ascending series of finite cyclic groups

$$\langle a_1 \rangle < \langle a_2 \rangle < \cdots < \langle a_n \rangle < \cdots,$$

where (in additive notation) $pa_1 = 0$, $pa_{n+1} = a_n$, $n = 1, 2, \ldots$. Consider the equation $sx = g$ where s is a positive integer and $g \in C_{p^\infty}$. The element g lies in some member of the series, say $\langle a_n \rangle$, i.e. $g = la_n$. Writing $s = p^k m$, $(m, p) = 1$, let d_1, d_2 be integers such that $1 = p^n d_1 + m d_2$. From this and the equations $g = la_n$, $p^n a_n = 0$, we get: $g = (p^n d_1 + m d_2)g = m d_2 la_n$. This together with the relation $p^k a_{n+k} = a_n$, yields in turn $g = mp^k (d_2 la_{n+k})$. Thus we have found a solution, namely $d_2 la_{n+k}$, for the equation $sx = g$.

The importance of these two examples derives from the fact that every divisible abelian group is a direct sum of isomorphic copies of Q and C_{p^∞}—see Theorem 9.1.6 below.

9.1.2. Exercise. Prove the divisibility of homomorphic images and direct and Cartesian products of divisible abelian groups.

9.1.3. Theorem. *Every abelian group can be embedded in some divisible abelian group.*

PROOF. Let G be an arbitrary abelian group and let F be a free abelian group with free generators x_i, $i \in I$, such that $G \simeq F/N$. Denote by F^* the direct sum $\bigoplus_{i \in I} Q_i$ of groups Q_i, isomorphic copies of the group Q of additive rationals, and choose from each summand Q_i an arbitrary nonzero element b_i. It is obvious that the map $x_i \to b_i$ extends to a monomorphism τ say, of the group F into the group F^*. On the strength of this monomorphism we may regard F as a subgroup of F^*. By the assertion contained in 9.1.2 above, the factor group F^*/N is divisible, and it also contains F/N ($\simeq G$) as a subgroup.

9.1.4. Theorem. *A divisible subgroup of an abelian group G is a direct summand of G.*

PROOF. Let A be a divisible subgroup of G, and denote by B a subgroup of G maximal with respect to having zero intersection with A (the existence of such a subgroup B follows from Zorn's lemma). We shall show that $G = A \oplus B$. (Obviously $\langle A, B \rangle = A \oplus B$.)

Suppose that G strictly contains $A \oplus B$, and let $g \in G \backslash (A \oplus B)$. The cyclic subgroup $\langle g \rangle$ has nonzero intersection with $A \oplus B$ since the contrary circumstance would imply that the sum $A + B + \langle g \rangle$ was direct, and then the subgroup $B \oplus \langle g \rangle$ would intersect A trivially, contradicting the choice of B. Let n be the smallest positive integer such that $ng \in A \oplus B$. We may suppose that n is prime, for if it is not we may consider $(n/p)g$ in place of g, where p is any prime divisor of n.

Let $ng = a + b$ where $a \in A$, $b \in B$. Being divisible, A contains an element a_1 such that $na_1 = a$. Replacing a by na_1 in the preceding equation we get $ng_1 = b$, where $g_1 = g - a_1$; of course g_1 is also outside $A \oplus B$.

By definition of B the intersection $A \cap \langle g_1, B \rangle$ is nonzero. This implies the existence of a nonzero element a' of A which can be written as a sum $a' = kg_1 + b'$, $b' \in B$, $0 < k < n$. Since $(k, n) = 1$, there exist integers l, s such that $lk + sn = 1$, so that $g_1 = lkg_1 + sng_1$. Since ng_1 and $kg_1 = a' - b'$ belong to $A \oplus B$, it follows that $g_1 \in A \oplus B$, which contradiction completes the proof.

9.1.5. Exercise. The sum (i.e. subgroup generated by) any set of divisible subgroups of an abelian group is again a divisible subgroup.

9.1.6. Theorem. *Every divisible group G decomposes as a direct sum of subgroups each isomorphic to either the additive group $\mathbf{Q}$ of rationals or a quasicyclic group $\mathbf{C}_{p^\infty}$ (where p may vary for different summands).*

PROOF. We obtain the desired decomposition by using transfinite induction. We may suppose G nonzero since the theorem is vacuously true otherwise. Choose in G any element $g \neq 0$. We consider two possibilities for g.

(i) If the element g has infinite order, then in view of the divisibility of G there is a sequence

$$g = g_1, g_2, \ldots, g_n, \ldots,$$

of elements of G, satisfying $(n + 1)g_{n+1} = g_n$, $n = 1, 2, \ldots$. It is easy to verify that the subgroup generated by all g_i is isomorphic to the group $\mathbf{Q}$ of additive rationals.

(ii) If g has finite order n say, then the element $a_1 = (n/p)g$, where p is any prime divisor of n, has order p. Again by the divisibility of G there is in G a sequence $a_1, a_2, \ldots$, of elements satisfying $pa_{n+1} = a_n$, and these generate a subgroup isomorphic to $\mathbf{C}_{p^\infty}$.

We have thus shown that our group G contains a subgroup A_1 isomorphic to either $\mathbf{Q}$ or $\mathbf{C}_{p^\infty}$.

Suppose we have defined an ascending series $A_1 < A_2 < \cdots < A_\beta < \cdots$ of divisible subgroups of G, indexed in order by the ordinals β less than α. (Recall that by the definition of ascending series for each limit ordinal β we have $A_\beta = \bigcup_{\delta < \beta} A_\delta$.) Suppose further that the above series has the property that for each β which is *not* a limit ordinal, the subgroup A_β is the direct sum of $A_{\beta-1}$ and a subgroup $C_{\beta-1}$ isomorphic to $\mathbf{Q}$ or some $\mathbf{C}_{p^\infty}$. If α is a limit ordinal we put $A_\alpha = \bigcup_{\beta < \alpha} A_\beta$. If α is not a limit ordinal and $A_{\alpha-1} \neq G$, then since $A_{\alpha-1}$ is divisible, by Theorem 9.1.4 we have that $G = A_{\alpha-1} \oplus C$, where C is nonzero and divisible. Just as we constructed the subgroup A_1 of the divisible group G, we construct in C a subgroup $C_{\alpha-1}$ isomorphic to $\mathbf{Q}$ or $\mathbf{C}_{p^\infty}$, and define $A_\alpha = A_{\alpha-1} \oplus C_{\alpha-1}$. In this way we define A_α inductively for all ordinals $\alpha \leq \gamma$, where γ is the first ordinal such that $A_\gamma = G$.

Having defined the A_α, $\alpha \le \gamma$, it only remains to observe that G is the direct sum of the subgroups $A_1 = C_0, C_1, \ldots, C_\alpha, \ldots$, for the proof to be complete.

9.1.7. Exercise. In a torsion-free abelian group the intersection of an arbitrary set of divisible subgroups is again divisible.

9.1.8. Exercise. A minimal divisible group G^* containing a given group G, is called a *divisible closure* of G. In other words a divisible closure G^* of the group is a divisible group containing G as a subgroup, such that if A is a divisible subgroup of G^* containing G then $A = G^*$. Prove that every torsion-free abelian group G has a torsion-free abelian divisible closure. Given any automorphism of G there is between any two such divisible closures of G an isomorphism extending the given automorphism.

9.1.9. Exercise. Show by means of an example that the intersection of divisible subgroups of a periodic group is not necessarily divisible.

9.1.10. Exercise. A group without nonzero divisible subgroups is said to be *reduced*. Every abelian group is the direct sum of a divisible subgroup and a reduced subgroup.

9.1.11. Exercise. In the theory of modules the concepts of projective and injective modules play important dual roles. In the case of abelian groups (regarded as modules over the ring of integers) these concepts may be defined as follows. An abelian group F is said to be *projective* (a *projective* **Z**-*module*), if for each *epi*morphism $\tau : A \to B$, and each homomorphism $\phi : F \to B$, (A, B abelian groups) there exists a homomorphism $\psi : F \to A$, such that the diagram

commutes, i.e. $\phi = \psi\tau$. An abelian group G is called *injective* if for each *mono*morphism τ from B to A and each homomorphism ϕ from B to G, there is a homomorphism ψ from A to G such that the diagram

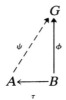

commutes, i.e. $\phi = \tau\psi$. Prove that an abelian group is (i) projective if and only if it is free, and (ii) injective if and only if it is divisible.

§10. Periodic Abelian Groups

In any abelian group G the set T of all elements of finite order is a subgroup; it is called the *torsion subgroup* of G. The factor group G/T is torsion-free. This fact to a certain extent reduces the study of arbitrary abelian groups to the study of periodic groups and torsion-free groups separately. It is however worth noting that in general the torsion subgroup is not a direct factor.

10.1.1. Example. Let

$$\hat{G} = \sum_p \mathbf{Z}_p, \qquad G = \bigoplus_p \mathbf{Z}_p,$$

(where p ranges over all primes). It is obvious that G is the torsion subgroup of $\hat{G}$. We shall show that $\hat{G}/G$ is divisible, and that G is not a direct summand of $\hat{G}$.

We first prove the divisibility of the quotient $\hat{G}/G$. Let $f \in \hat{G}$, and let n be a positive integer. Since for $p > n$ the group $\mathbf{Z}_p$ contains an element g_p satisfying $ng_p = f(p)$, we have that $ng = f'$, where g and f' are defined as follows:

$$g(p) = \begin{cases} 0 & \text{for } p \leq n, \\ g_p & \text{for } p > n; \end{cases} \qquad f'(p) = \begin{cases} 0 & \text{for } p \leq n, \\ f(p) & \text{for } p > n. \end{cases}$$

Then since clearly $Gf = Gf'$, it follows that the equation $nx = Gf$ has a solution in $\hat{G}/G$ (Gg will do).

Now suppose that G is a direct summand of $\hat{G}$, say $\hat{G} = G \oplus H$. It follows from the above that since $H \simeq \hat{G}/G$, H is divisible. Thus for every positive integer n and any $h \in H$, the equation $nx = h$ has a solution in H. However if h is such that $h(p) \neq 0$ then this equation has no solution if $n = p$. This contradiction proves our assertion about G.

In a periodic abelian group G the set G_p of all p-elements, i.e. elements whose orders are powers of the fixed prime p, forms a subgroup. It is obvious that G_p is the unique maximal p-subgroup of G; it is also called the *p-primary component* of the group G.

10.1.2. Exercise. A periodic abelian group is the direct sum of its primary components.

In view of this statement, in studying periodic abelian groups we may—at least insofar as we are considering questions of decomposability—restrict attention to p-groups. We shall therefore limit ourselves in this section to stating and proving results about p-groups only; the appropriate generalizations to arbitrary periodic abelian groups may be deduced easily.

The theorem about the decomposability of a finitely generated abelian group as a direct sum of cyclic subgroups naturally prompts the question of the existence of such a decomposition for an arbitrary abelian p-group without divisible subgroups. In general the answer is in the negative; however there are useful conditions under which an abelian p-group *is* a direct sum of cyclic subgroups.

10.1.3. EXAMPLE. Let $G = \bigoplus_n \langle a_n \rangle$ be the direct sum of its cyclic subgroups $\langle a_n \rangle$ of orders $|a_n| = p^n$, $n = 1, 2, \ldots$, and write $b_n = p^{n-1} a_n$. Put $N = \langle c_1, \ldots, c_n, \ldots \rangle$, where $c_n = b_n - b_{n+1}$. We shall prove that the group $\hat{G} = G/N$ has no nontrivial divisible subgroups, and does not decompose as a direct sum of cyclic subgroups.

To see that $\hat{G}$ has no nontrivial divisible subgroups write $N_1 = \langle b_1, N \rangle$; then $\hat{N}_1 = N_1/N$ is finite, and the quotient $\hat{G}/\hat{N}_1$ is the direct sum of its cyclic subgroups $\langle \hat{a}_n, \hat{N}_1 \rangle / \hat{N}_1$, where $\hat{a}_n = a_n + N$. It is easy to show that such groups have no nonzero divisible subgroups, and thence that $\hat{G}$ also has no nonzero divisible subgroups.

We now prove the indecomposability of $\hat{G}$ as a direct sum of cyclic subgroups. Since $b_i + N = b_{i+1} + N$, we have that $p^{n-1} \hat{a}_n = \hat{a}_1$, so that for every $n > 0$ (and the fixed element $\hat{a}_1$) the equation $p^n x = \hat{a}_1$ has a solution in $\hat{G}$. However it is obvious that a direct sum of cyclic p-groups can never have this property, giving us the desired conclusion.

Recall that the exponent of a p-group G is that integer p^n (if it exists) for which $p^n G = 0$, while $p^{n-1} G \neq 0$.

10.1.4. Exercise. An abelian group of prime exponent p is a direct sum of cyclic subgroups. (Hint. Use transfinite induction, or regard the abelian group of exponent p as a vector space over the field $\mathbf{GF}(p)$ and apply the theorem from linear algebra about the decomposition of a vector space as a direct sum of one-dimensional subspaces.)

10.1.5. Theorem (Prüfer's First Theorem). *An abelian p-group of finite exponent is a direct sum of cyclic subgroups.*

PROOF. We shall use induction on n (starting from $n = 1$), where the exponent of the group is p^n. On the strength of Exercise 10.1.4 above we may proceed immediately to the inductive step: assume that the theorem is true for groups of exponent p^k, $k < n$, and consider an abelian group G of exponent p^n.

Since the subgroup pG has exponent p^{n-1}, it is by the inductive hypothesis a direct sum of cyclic subgroups: $pG = \bigoplus_{i \in I} \langle a_i \rangle$. Denote by x_i some solution (in G) of the equation $px = a_i$ (a solution exists in G since $a_i \in pG$), and put $H = \langle x_i \mid i \in I \rangle$. Then $H = \bigoplus_{i \in I} \langle x_i \rangle$ (prove it!). Let B be a subgroup of G maximal with respect to intersecting H trivially, and suppose $G \neq H + B$. Let $g \in G \backslash (H + B)$. By construction of the subgroup H it is clear that the equation $px = pg$ has at least one solution in H, say h. Then the element $g_1 = g - h$, which does not lie in $H + B$, satisfies $pg_1 = 0$. By the definition of B, the intersection $H \cap \langle g_1, B \rangle$ is nonzero. This means that there is a nonzero element $h_1 \in H$ which can be written in the form $h_1 = kg_1 + b$, $b \in B$, $0 < k < p$. Hence if $sk \equiv 1 \pmod{p}$, then $g_1 = skg_1 = sh_1 - sb \in H + B$, a contradiction. Thus $G = H \oplus B$.

Since the subgroup H (by construction), and the subgroup B (by 10.1.4), decompose as direct sums of cyclic subgroups, so also does G. This completes the proof.

We shall say that an element $g \neq 0$ of an abelian p-group G has *finite height* h in G, if the equation $p^n x = g$ has a solution only for $n \leq h$. If the equation $p^n x = g$ has a solution in G for every n, the height of g will be defined to be infinite.

In terms of the concept of height we can give a further sufficient condition for the decomposability of an abelian p-group as a direct sum of cyclic subgroups. However before stating this condition we introduce the useful idea of a pure subgroup.

A subgroup A of a group G is said to be *pure* if for every integer n and every element $a \in A$, whenever the equation $nx = a$ is soluble in G, then it is soluble in the subgroup A.

10.1.6. Exercise. In a direct sum of abelian groups each summand is pure.

10.1.7. Exercise. The torsion subgroup of an abelian group G is pure in G.

10.1.8. Exercise. A subgroup A of an abelian p-group G is pure if and only if whenever an equation of the form $p^n x = a$, $a \in A$, has a solution in G then it has a solution in A.

10.1.9. Exercise. In an abelian group G satisfying $nG = 0$, a subgroup A is pure if and only if for every divisor m of n and every $a \in A$, the solubility of the equation $mx = a$ in G, implies its solubility in the subgroup A.

10.1.10. Exercise. Suppose A is a pure subgroup of an abelian group G such that the quotient of G by A is cyclic, say $G/A = \langle g + A \rangle$. Then in the coset $g + A$ there is an element g_1 such that $|g_1| = |G/A|$.

10.1.11. Exercise. If the quotient of an abelian group G by a pure subgroup A decomposes as a direct sum of cyclic subgroups, then A is a direct summand of G. (Hint. Use Exercise 10.1.10.)

10.1.12. Theorem (L. Ja. Kulikov-Prüfer). *If A is a pure subgroup of finite exponent of an abelian group G, then A is a direct summand of G.*

PROOF. By hypothesis there is a positive integer n such that $nA = 0$. This and the purity of A imply that $A \cap nG = 0$.

We shall establish the purity of the subgroup B/nG (where $B = A + nG$) in its containing group G/nG. Thus suppose that the equation $mx = a + nG$, $a \in A$, has the solution $g + nG$ say, in G/nG. By 10.1.9 we may assume that m divides n, say $n = mm_1$.

From the relation $m(g + nG) = a + nG$ we infer that $mg = a + ng_1$, $g_1 \in G$, and thence that $a = m(g - m_1 g_1)$. From this and the purity of A we get the existence in A of an element a_1 such that $a = ma_1$; but then $a_1 + nG$ is a solution in B/nG of our original equation $mx = a + nG$.

Since B/nG is pure in G/nG, and by Prüfer's First Theorem the quotient group G/nG is a direct sum of cyclic subgroups, we have by 10.1.11 that its subgroup B/nG is a direct summand: $G/nG = B/nG \oplus C/nG$ say. From this and $A \cap nG = 0$, we deduce that $G = A \oplus C$, completing the proof.

10.1.13. Corollary. *If the torsion subgroup T of an abelian group G has finite exponent then T is a direct summand of G.*

10.1.14. Theorem (Prüfer's Second Theorem). *A countable abelian p-group without elements of infinite height is a direct sum of cyclic subgroups.*

PROOF. Let G be a group satisfying the hypothesis. Since G is countable so is the subgroup $A = \{g \mid g \in G, pg = 0\}$ (the "bottom layer" of G). Hence there exists a chain

$$0 = A_0 < A_1 < \cdots < A_n < \cdots$$

of subgroups such that $\bigcup A_n = A$ and $|A_{n+1} : A_n| = p$.

Let $A_1 = \langle a_1 \rangle$ and let b_1 be a solution of the equation $p^{h_1} x = a_1$, where h_1 is the height of the element a_1. It is easy to see that the subgroup $\langle b_1 \rangle$ is pure in G, whence $\langle b_1 \rangle$ is a direct summand of G, say $G = \langle b_1 \rangle \oplus B_1$. It is then obvious that $A_2 = A_1 + (A_2 \cap B_1)$, and that every nonzero element $a_2 \in A_2 \cap B_1$ has the same height, h_2 say, in G as in B_1.

Denote by $b_2 \in B_1$ any solution of the equation $p^{h_2} x = a_2$. As before there is a direct decomposition $B_1 = \langle b_2 \rangle \oplus B_2$ say, and then $G = \langle b_1 \rangle \oplus \langle b_2 \rangle \oplus B_2$.

Continuing this process we obtain for all positive integers n subgroups $\langle b_n \rangle$ and B_n with properties that the $\langle b_n \rangle$ generate their direct sum C say: $C = \langle b_1 \rangle \oplus \langle b_2 \rangle \oplus \cdots$, the B_n form a chain $B_1 > B_2 > \cdots$, $G = \langle b_1 \rangle \oplus \cdots \oplus \langle b_n \rangle \oplus B_n$, and finally $B_n \cap A_n = 0$.

We prove now that $G = C$. Let $0 \neq g \in G$, $|g| = p^m$. Choose an n such that $p^{m-1} g \in A_n$, and write $g = c + b$ where $c \in \langle b_1 \rangle \oplus \cdots \oplus \langle b_n \rangle$, $b \in B_n$. Since $p^{m-1} b \in A_n \cap B_n = 0$, we have $|b| < |g|$ so that, using induction, we may assume that $b \in C$. But then $g \in C$, and the theorem is proved.

The following example shows that the countability assumption cannot be removed.

10.1.15. EXAMPLE. Let p be a prime, $G = \sum_n \mathbf{Z}_{p^n}$, and let T be the torsion subgroup of G. Clearly T is a p-group without elements of infinite height and has the cardinal of the continuum. We show that T does not decompose as the direct sum of cyclic subgroups. Suppose on the contrary that $T = \bigoplus_{i \in I} \langle a_i \rangle$, $a_i \neq 0$. Since T is uncountable we can find an infinite subset $I_1 \subseteq I$ such that the subgroup $A = \bigoplus_{i \in I_1} \langle a_i \rangle$ has finite exponent, say $p^k A = 0$. Since A is a direct summand of T, the heights in T of the non-zero elements of A cannot exceed k. Thus if b_n generates $\mathbf{Z}_{p^n}$, then for every $f \in A$ we have

$$f(n) \in \langle p^{n-k} b_n \rangle \quad \text{for } n \geq k.$$

Since the nth components of functions $f \in G$ can take only finitely many values (p^n to be precise), and the subgroup A is infinite, it follows that in A there are distinct elements f_1, f_2 with the same first $2k$ components; i.e. $f_1(n) = f_2(n)$, $1 \leq n \leq 2k$. (There are only finitely many projections of the elements of A on the first r components for any fixed finite r.) Their difference $f_1 - f_2$ is nonzero, lies in A, and has height exceeding k in T (since the first $2k$ components are zero and for $n > 2k$ the nth component is a multiple of $p^{n-k} b_n$), contradicting the definition of A.

Finite Groups 4

§11. Sylow p-Subgroups

In the preceding chapter we saw that the structure of an abelian group is largely determined by the structure of its maximal p-subgroups. In the theory of finite groups maximal p-subgroups play a similarly crucial role. In this section we shall prove the following theorem of Sylow about finite groups: For every prime power p^α dividing the order of a finite group, there is a subgroup of order p^α; if $p^{\alpha+1}$ divides the order of the group then each subgroup of order p^α is contained in some subgroup of order $p^{\alpha+1}$; all maximal p-subgroups are conjugate in the group, and the number of such subgroups is congruent to 1 modulo p. This theorem, first proved by the Norwegian mathematician L. Sylow just over a century ago (in 1872), has turned out to be the cornerstone of finite group theory. It has since been generalized in several different directions both in the Soviet Union (by S. A. Čunihin and others), and elsewhere (by P. Hall and others). Because of the theorem's importance, in honor of its discoverer the maximal p-subgroups of a finite group (and for that matter of an infinite group) are more often called *Sylow p-subgroups*.

From Sylow's theorem it follows in particular that the Sylow p-subgroups of a finite group are just the subgroups of order p^r, where p^r is the largest power of p dividing the group order. We note that if the positive integer m divides the order of a finite group G but is not a prime power, then G need not in general have a subgroup of order m—for instance the alternating group $\mathbf{A}_4$, which has order 12, has no subgroup of order 6; see Exercise 11.2.2 below.

11.1. Sylow's Theorem

For the proof of Sylow's theorem we shall find very useful the idea of the action of a group on a set. We say that a group G *acts on a set* $\mathcal{M}$, if for each $m \in \mathcal{M}$, $g \in G$, and element $mg \in \mathcal{M}$ is defined such that $(mg_1)g_2 = m(g_1 g_2)$ and $me = m$ for all $m \in \mathcal{M}$, $g_1, g_2 \in G$ (where e is of course the identity element). In other words, G acts on $\mathcal{M}$ if G comes equipped with a homomorphism $\phi: G \to S(\mathcal{M})$ (starting from this point of view mg is then defined as $m(g^\phi)$). The set $mG = \{mg \mid g \in G\}$ is called the *orbit* of the element m (or the orbit of G (under the given action) containing m). Clearly the orbits of two elements of $\mathcal{M}$ either coincide or have empty intersection, so that the orbits form a partition of $\mathcal{M}$.

11.1.1. Theorem (Sylow). *Let G be a finite group and let p be a prime. EXISTENCE: For each power p^α dividing the order of G there is a subgroup of G of order p^α. INCLUSION: If $p^{\alpha+1}$ divides the order of G then for each subgroup of G of order p^α there is one of order $p^{\alpha+1}$ containing it. In particular the Sylow p-subgroups of G are just the subgroups of order p^r where p^r is the largest power of p dividing the order of G. CONJUGACY: The Sylow p-subgroups of G form a single conjugacy class of subgroups of G. NUMBER: The number of Sylow p-subgroups of G is congruent to 1 modulo p, and divides the order of G.*

PROOF (H. Wielandt). *Existence.* Let $|G| = p^r l$, $(p, l) = 1$, and let $\mathcal{M}$ be the set of all subsets of G of cardinal p^α. Obviously

$$|\mathcal{M}| = \binom{p^r l}{p^\alpha} = p^{r-\alpha} l \prod_{j=1}^{p^\alpha - 1} \frac{p^r l - j}{j},$$

so that the largest power of p dividing $|\mathcal{M}|$ is $p^{r-\alpha}$. If $M \in \mathcal{M}$, $g \in G$, then clearly $Mg = \{mg \mid m \in M\} \in \mathcal{M}$, so that G acts on $\mathcal{M}$ by multiplication on the right. Let $\{M_1, \ldots, M_s\}$ be an orbit whose size s is not divisible by $p^{r-\alpha+1}$, and further write

$$G_i = \{g \mid g \in G, M_1 g = M_i\}, \qquad 1 \leq i \leq s.$$

It is readily verified that G_1 is a subgroup of G and that the G_i are the right cosets of G_1 in G. We shall show that G_1 has the required order p^α. For the time being write $|G_1| = t$; then by Lagrange's theorem (2.4.5) $st = |G| = p^r l$. Since $s \nmid p^{r-\alpha+1}$, we must have that $p^\alpha | t$, so that certainly $t \geq p^\alpha$. On the other hand for $x \in M_1$ we obviously have that $xG_1 \subseteq M_1$, whence $|G_1| \leq |M_1|$, or $t \leq p^\alpha$; therefore $t = p^\alpha$.

Inclusion. Suppose $p^{\alpha+1}$ divides $|G|$. Let P be a subgroup of G of order p^α, and let $\mathcal{P}$ be the class of subgroups of G conjugate to P. We know that

$$|\mathcal{P}| = |G : N_G(P)|.$$

If p does not divide $|\mathcal{P}|$ then this equality tells us that $p^{\alpha+1}$ must divide

$N_G(P)$, so that by the first part of the theorem the group $N_G(P)/P$ contains a subgroup P^*/P say, of order p. Then P^* will serve as the required super-group of P. Suppose now that p does divide $|\mathscr{P}|$. If we let the group P act on $\mathscr{P}$ by conjugation, then the orbits of P under this action have orders dividing $|P|$, and so of the form p^{α_i}, $\alpha_i \geq 0$. Since there is at least one one-element orbit—we have $\{P\}$ in mind here—and p divides $|\mathscr{P}|$, there must be at least one other one-element orbit, $\{Q\}$ say. But this means that P normalizes Q, whence PQ is a p-subgroup (since $PQ/Q \simeq P/P \cap Q$, and the extension of a p-group by a p-group is again a p-group). Applying to PQ that inner automorphism of G which maps Q onto P, we obtain a p-subgroup $\hat{P}P$ containing P as a normal subgroup. Again by the first part of the theorem there is in $\hat{P}P/P$ a subgroup P^*/P or order p, and then P^* is the desired containing group.

As remarked previously, it now follows that the Sylow p-subgroups of a finite group are precisely the subgroups of order p^r where p^r is the largest power of p dividing the order of the group.

Conjugacy. Now let P be a Sylow p-subgroup of G, so that then $|P| = p^r$, and let $\mathscr{P}$ be defined as above in terms of this P. We wish to show that every Sylow p-subgroup of G lies in $\mathscr{P}$. Thus let Q be any Sylow p-subgroup of G; then if we think of Q as acting on $\mathscr{P}$ by conjugation, the orbit-sizes of Q under this action must all divide $|Q|$, and must therefore be powers of p. Since p does not divide $|\mathscr{P}|$ (using $|\mathscr{P}| = |G:N_G(P)|$ and that P is a maximal p-subgroup), we have as before that at least one of these orbits must be a singleton, say $\{\hat{P}\}$, and then Q normalizes $\hat{P}$. But then $\hat{P}Q$ is a p-subgroup, and the maximality of $\hat{P}$, Q gives that $Q = \hat{P}Q = \hat{P} \in \mathscr{P}$.

Number. In the notation of the preceding paragraph, it suffices to prove that $\{Q\}$ is the only singleton orbit. This is easy, since if $\{\hat{Q}\}$ were another such orbit, then $Q\hat{Q}$ would be a p-subgroup properly containing Q, contradicting the maximality of Q. This completes the proof of the theorem.

It is perhaps more traditional to speak of the *three* Sylow theorems rather than just one: the assertions about existence and inclusion are some-times grouped together as the first Sylow theorem, the assertions about conjugacy and number form respectively the second and third Sylow theorems.

11.1.2. Exercise. A group of order 196 contains a normal Sylow p-sub-group.

11.1.3. Exercise. The Sylow p-subgroups of an infinite group are not always conjugate. (Hence the assumption of finiteness in Sylow's theorem is essential.) (Hint. Consider an infinite direct power of S_3.)

11.1.4. Exercise. Let P be a Sylow p-subgroup of a finite group G, and let H be a subgroup containing the normalizer $N_G(P)$ of P. Then $N_G(H) = H$.

11.2. An Application to Groups of Order pq

Sylow's theorem often allows us to obtain a lot of information about a given group, and even to describe completely the structure of groups which, in some sense or other, have small orders. By way of illustration we shall now analyse the structure of the groups of order pq.

Let p, q be primes with $p < q$. What must a group G of order pq look like? The Sylow p- and q-subgroups are of prime orders and so cyclic. Let $\langle a \rangle$, $\langle b \rangle$ be respectively Sylow p- and q-subgroups. By Sylow's theorem the number of Sylow q-subgroups of G has the form $1 + kq$, and divides pq, so that there can only be one of them; i.e. $\langle b \rangle$ is the unique Sylow q-subgroup of G. In particular it follows that $\langle b \rangle$ is normal in G. The number of Sylow p-subgroups of G has the form $1 + kp$ and divides q; this gives us two possibilities:

(i) The Sylow p-subgroup $\langle a \rangle$ is unique. Then it is normal, and since $\langle a \rangle \cap \langle b \rangle = 1$ we have that $G = \langle a \rangle \times \langle b \rangle$. Since ab has order pq it follows that $G = \langle ab \rangle \simeq \mathbf{Z}_{pq}$.

(ii) There are exactly q Sylow p-subgroups. Of course this can happen only if $q \equiv 1 \pmod{p}$. Suppse $a^{-1}ba = b^r$. We cannot have $r = 1$ for then we would be in Case (i). By successive conjugation we get that $a^{-m}ba^m = b^{r^m}$, whence

$$a^{-m}b^n a^m = b^{r^m n}$$

for all positive integral m, n. For $m = p, n = 1$ this gives $r^p \equiv 1 \pmod{q}$, and it also gives a formula for multiplication:

$$a^m b^n \cdot a^s b^t = a^{m+s} b^{nr^s + t}. \tag{1}$$

Conversely, it is easy to verify that if $q \equiv 1 \pmod{p}$, $r^p \equiv 1 \pmod{q}$, $r \not\equiv 1 \pmod{q}$, then this multiplication formula for expressions of the form $a^i b^j$, defines a nonabelian group of order pq. Finally, the solutions of the congruence $x^p \equiv 1 \pmod{q}$ constitute a cyclic group of order p, so that those solutions $\not\equiv 1 \pmod{q}$ have the form $r, r^2, \ldots, r^{p-1}$, where r is any one of them. If we take any of these solutions in place of the particular one r appearing in (1), we get the same group, since replacing r by r^i amounts to the same thing as taking a^i in place of a.

With the help of Sylow's theorem we have thus described all possible groups of order pq; it turns out that there are (up to isomorphism) at most two, one abelian and one nonabelian; the nonabelian one exists precisely if $q \equiv 1 \pmod{p}$.

11.2.1. Exercise. There are exactly two nonisomorphic groups of order 6: the cyclic group $\mathbf{Z}_6$ and the symmetric group $\mathbf{S}_3$.

11.2.2. Exercise. The alternating group $\mathbf{A}_4$ does not contain a subgroup of order 6, although 6 divides its order (12). Hence the converse of Lagrange's theorem is false in general. (This fact was noted also at the beginning of this section.)

(Solution. By Exercise 11.2.1 if A_4 contained a subgroup of order 6, then this subgroup would have to be isomorphic to Z_6 or S_3. Since it is obvious that no permutation of 4 letters can have order 6, the first possibility is ruled out. The second is ruled out by the fact that in S_3 there are 3 elements of order 2, and the same is true of A_4—the elements of order 2 in A_4 are (12)(34), (13)(24), (14)(23). Thus these elements would all have to lie in the subgroup; however they generate a group of order 4, and of course S_3 can have no such subgroup.)

11.3. Examples of Sylow *p*-Subgroups

We turn our attention again to the groups of Examples (I)–(IV) of Chapter 1.

11.3.1. EXAMPLES. (I) The group Z_n decomposes as the direct sum of its Sylow *p*-subgroups, which are cyclic of orders $p_1^{m_1}, \ldots, p_s^{m_s}$, where n has the prime decomposition $n = p_1^{m_1} \cdots p_s^{m_s}$.

(II). The Sylow *p*-subgroup of the multiplicative group $\mathbf{C}^*$ is the quasi-cyclic group C_{p^∞}.

(III). We next describe (following Kalužnin) the Sylow *p*-subgroups of the symmetric groups. Since $|S_n| = n!$, it is appropriate to ask for those exponents $e(n)$ such that $p^{e(n)}$ divides $n!$. The multiples of p among the numbers $1, 2, \ldots, n$ are the numbers $p, 2p, \ldots, kp$, where $k = [n/p]$ (and $[n/p]$ denotes the largest integer $\leq n/p$), so that $e(n) = [n/p] + e(k)$. Since $[k/p] = [n/p^2]$, we have that

$$e(n) = \left[\frac{n}{p}\right] + \left[\frac{n}{p^2}\right] + \cdots.$$

If we represent n as a sum of powers of p, say

$$n = a_0 + a_1 p + \cdots + a_s p^s, \qquad 0 \leq a_i < p, \tag{2}$$

then

$$e(n) = a_1 + a_2(1+p) + a_3(1+p+p^2) + \cdots + a_s(1+p+\cdots+p^{s-1}). \tag{3}$$

We consider the group S_n first in the case that n is a power of p. Suppose that in S_{p^m} we have already formed a Sylow *p*-subgroup; i.e. a subgroup H_m of order $p^{(1+\cdots+p^{m-1})}$. Using this subgroup we shall construct a subgroup H_{m+1} of $S_{p^{m+1}}$ of order $p^{(1+\cdots+p^m)}$. To this end we split the sequence $1, 2, \ldots, p^{m+1}$ of permuted symbols into p subsequences, or segments, of consecutive symbols, each segment of length p^m. If we set

$$c = \prod_{j=1}^{p^m} (j, p^m + j, 2p^m + j, \ldots, (p-1)p^m + j),$$

then if x is a permutation moving only the symbols of the ith segment, it is easy to see that $c^{-1}xc$ moves only the symbols of the $(i+1)$st segment (where $i+1$ is reduced modulo p if necessary). It follows that the subgroup of $\mathbf{S}_{p^{m+1}}$ generated by the subgroups $c^{-r}H_m c^r$, $0 \le r < p$, is the direct product of these subgroups, so that the subgroup generated by the subgroup H_m and the element c is isomorphic to the wreath product $H_m \operatorname{wr}\langle c \rangle$. This is the desired H_{m+1}, since

$$|H_m \operatorname{wr}\langle c \rangle| = |H_m|^p |c| = p^{(1+\cdots+p^m)}.$$

It now becomes clear that each Sylow p-subgroup of $\mathbf{S}_{p^m}$ is isomorphic to the group

$$(\cdots (\mathbf{Z}_p \operatorname{wr} \mathbf{Z}_p) \operatorname{wr} \cdots) \operatorname{wr} \mathbf{Z}_p$$

obtained by wreathing with the cyclic group $\mathbf{Z}_p$ m times.

Now let n be arbitrary. We partition the sequence $1, 2, \ldots, n$, into a_0 one-element segments, a_1 p-element segments, and so on (see (2)). The symmetric group on each of these segments will be of degree p^m for some m. In each of these symmetric groups choose a Sylow p-subgroup (constructed above). Since these subgroups act on pairwise disjoint sets of symbols, the group they generate, call it P_n, will be their direct product, and will therefore have order

$$P_n = \prod_{m=1}^{s} p^{(1+p+\cdots+p^{m-1})a_m} = p^{e(n)}$$

(see (3)). Hence P_n is a Sylow p-subgroup of $\mathbf{S}_n$. We have thus shown that P_n is isomorphic to the direct product of several iterated wreath products of the form $(\cdots (\mathbf{Z}_p \operatorname{wr} \mathbf{Z}_p) \operatorname{wr} \cdots) \operatorname{wr} \mathbf{Z}_p$.

(IV). Finally we consider the general linear group over finite fields. Let p be a prime, m, n integers ≥ 1, and $q = p^m$. We shall show that $\mathbf{UT}_n(q)$ is a Sylow p-subgroup of the group $\mathbf{GL}_n(q)$—simply by calculating their orders.

Which n-vectors over $\mathbf{GF}(q)$ can be the first row of a nonsingular matrix? The answer is clear: any but the zero vector; thus there are $q^n - 1$ possibilities for the first row. Once the first row is chosen then for the second any n-vector not a multiple of the first will do; there are $q^n - q$ such vectors. Given the first two rows, then as the third we may take any n-vector linearly independent of the first two; there are thus $q^n - q^2$ possibilities. Continuing in this way, we arrive at the formula

$$|\mathbf{GL}_n(q)| = \prod_{i=0}^{n-1} (q^n - q^i). \tag{4}$$

Since the entries above the main diagonal in matrices from $\mathbf{UT}_n(q)$ range independently of one another over the whole field, and since each matrix has $\binom{n}{2}$ entries above its main diagonal, we have that

$$|\mathbf{UT}_n(q)| = q^{\binom{n}{2}}.$$

By a comparison of orders we see that $\mathbf{UT}_n(q)$ is a Sylow p-subgroup of $\mathbf{GL}_n(q)$.

11.3.2. Exercise. Write down permutations in S_4 forming a Sylow 2-subgroup of S_4.

11.3.3. Exercise. Let p be a prime, m and n positive integers, and let $\mathbf{P}_n(\mathbf{Z}_{p^m})$ denote the set of all those matrices of $\mathbf{GL}_n(\mathbf{Z}_{p^m})$ with entries under the main diagonal all multiples of p and with diagonal entries all congruent to 1 modulo p. Verify that $\mathbf{P}_n(\mathbf{Z}_{p^m})$ is a Sylow p-subgroup of $\mathbf{GL}_n(\mathbf{Z}_{p^m})$. (Hint. With the aid of the homomorphism $\mathbf{GL}_n(\mathbf{Z}_{p^m}) \to \mathbf{GL}_n(\mathbf{Z}_p)$ derive the formula

$$|\mathbf{GL}_n(\mathbf{Z}_{p^m})| = \prod_{i=0}^{n-1} (p^{mn} - p^{(mn-n+i)}).) \tag{5}$$

To conclude this section we state in the form of exercises the following few observations.

11.3.4. Exercise. Let ϕ be a homomorphism of a finite group G. If P is a Sylow p-subgroup of G, then P^ϕ will be a Sylow p-subgroup of G^ϕ. Conversely every Sylow p-subgroup of G^ϕ is the image of some Sylow p-subgroup of G. This remark is sometimes useful in proofs by induction.

11.3.5. Exercise. It can happen that the product of two elements of order p has finite order coprime to p of even infinite order. Thus the elements of order p^α, p fixed, by no means always form a group.

11.3.6. Exercise. Let G be a finite group and let $A \le G$. If the index $|G:A|$ is less than some prime divisor p of the order of the group G, then the intersection $\bigcap_{g \in G} A^g$ contains a Sylow p-subgroup of G (and so, in particular, is nontrivial).

§12. Finite Simple Groups

Just as the natural numbers are built up from the prime numbers by means of multiplication, so finite groups are built up from the finite simple groups by means of extensions. For, consider a composition series

$$1 = G_0 < G_1 < \cdots < G_m = G$$

for a finite group G (i.e., as defined in Section 4.4, Chapter 2, a finite sequence of subgroups beginning with 1 and ending with G in which every member is maximal normal in its successor—clearly every finite group has such a series). If some factor G_{i+1}/G_i were not simple then by taking a proper normal subgroup H/G_i of it, we could refine the series by inserting H between G_i and G_{i+1}; hence the factors of a composition series are always simple. It is in this sense that we talk of the finite simple groups as being the building blocks of finite groups, although, as opposed to the situation for numbers, the end result is not determined solely by the blocks entering into its construction, but depends also on the way they are piled on top of one another.

The most obvious examples of simple groups are the cyclic groups of prime orders; it is obvious that these are the only abelian simple groups. The classification of all finite simple groups is an extremely important, indeed at the present time perhaps the main, problem in the theory of finite groups (although results pertaining to it do not exhaust the whole theory). This problem is being attacked on several fronts: one school develops general "industrial" methods, aiming at discovering how to obtain by a single procedure all finite simple groups—to this school belong also deep results concerning the identification of groups obtained in various ways; a second school, on the other hand, pinning its hopes on intuition and assiduity, searches for sporadic examples—at the present time the value of each new finite simple group is unusually high; a third school attempts to classify all finite simple groups with one or another property—for example with a given Sylow p-subgroup, etc. Not long ago a famous fifty-year old problem of Burnside was solved—W. Feit and J. G. Thompson [Solvability of groups of odd order, Pacific J. Math. **13** (1963), 775–1029] proved that every nonabelian finite simple group has even order.

The aim of the present section is a comparatively modest one: to give two of the classical series of finite simple groups, namely the alternating groups and the projective special linear groups. The proofs which follow are tailored to our needs and do not give any idea of the above-mentioned industrial methods. A good summary of the present state of the subject and tables of the known simple groups can be found in [6, 24, 28].

12.1. The Alternating Groups

In this section the following facts from Chapter 1 will be used repeatedly: the formula

$$|x^G| = |G:N_G(x)|,$$

and the result that the permutations with the same disjoint cycle decompositions (i.e. the same number of cycles of each order) form a single conjugacy class in $\mathbf{S}_n$, while in $\mathbf{A}_n$ they may comprise either a single conjugacy class, or else two conjugacy classes of the same size.

12.1.1. Theorem (Galois). *For $n \neq 4$ the alternating group $\mathbf{A}_n$ is simple. The group $\mathbf{A}_4$ is not simple.*

PROOF. We shall examine the groups $\mathbf{A}_n$ on an individual basis for $n < 7$, while for $n \geq 7$ we shall give a general proof of simplicity. Thus although our proof will not be the shortest possible, it will enable us to become more familiar with the alternating groups of small degree.

The groups $\mathbf{A}_1$, $\mathbf{A}_2$, $\mathbf{A}_3$ have orders 1, 2, 3 respectively and are therefore simple.

The group $\mathbf{A}_4$ is not simple: it has a "polycyclic" composition series with factors of orders 2, 2, 3—see 4.4.1 (III).

To show that $\mathbf{A}_5$ is simple we shall calculate its conjugacy classes, and then verify that no union of conjugacy classes yields a proper subgroup. As the first step we shall consider the types of possible disjoint cycle decompositions of elements of $\mathbf{A}_5$.

(i) Elements of type $x = (123)$: In the group $\mathbf{S}_5$, and therefore also in $\mathbf{A}_5$, the number of such elements is $\frac{1}{3} \cdot 5 \cdot 4 \cdot 3 = 20$. They are of course all conjugate in $\mathbf{S}_5$, so that $|\mathbf{S}_5 : N_{\mathbf{S}_5}(x)| = 20$. Since $|\mathbf{S}_5| = 120$, we have that $|N_{\mathbf{S}_5}(x)| = 6$. It is clear that the powers of x and the transposition (45) normalize x, so that

$$N_{\mathbf{S}_5}(x) = \{1, (123), (132), (45), (123)(45), (132)(45)\}.$$

Since three of these six permutations are odd, we get that $N_{\mathbf{A}_5}(x) = 3$, whence $|\mathbf{A}_5 : N_{\mathbf{A}_5}(x)| = 20$, which in turn implies that the cycles of type (123) form a single conjugacy class of $\mathbf{A}_5$.

(ii) Elements of type $y = (12)(34)$: Altogether there are $(1/2) \cdot (5 \cdot 4/2) \cdot (3 \cdot 2/2) = 15$ such items. Since they cannot be partitioned into two sets of equal size, they must form a single conjugacy class of $\mathbf{A}_5$.

(iii) Elements of type $z = (12345)$: there are altogether $\frac{1}{5} \cdot 5! = 24$ of them, so that $|N_{\mathbf{S}_5}(z)| = 120/24 = 5$, i.e. the normalizer of z consists of just the powers of z. Since these are all even permutations, we get that $|N_{\mathbf{A}_5}(z)| = 5$, and thence that in $\mathbf{A}_5$ there are just $60/5 = 12$ elements conjugate to z. Thus the set of cycles of length 5 splits into two conjugacy classes.

From these calculations we deduce that if H is a normal subgroup of $\mathbf{A}_5$, then

$$|H| = 1 + 20r + 15s + 12t, \quad \text{where } 0 \leq r \leq 1, 0 \leq s \leq 1, 0 \leq t \leq 2.$$

By considering the twelve possibilities for the triple r, s, t, and taking into account that $|H|$ must divide $|\mathbf{A}_4| = 60$, we conclude that $|H| = 1$ or 60. This proves that $\mathbf{A}_5$ is simple.

For the remainder of the proof we shall make use of the following remark:

(iv) If $n \geq 5$ and H is a normal subgroup of $\mathbf{A}_n$ containing a 3-cycle, then $H = \mathbf{A}_n$. To see this first observe that since the group $\mathbf{A}_n$ is generated by the 3-cycles, it suffices to show that an arbitrary 3-cycle (ijk) is conjugate in $\mathbf{A}_n$ to (123). If these two cycles have at least one symbol in common, then together they move at most 5 symbols, so that they lie in the subgroup A say, of $\mathbf{A}_n$, fixing a certain $(n-5)$ symbols. Obviously $A \approx \mathbf{A}_5$; therefore the cycles (ijk) and (123) are already conjugate in A (see (i) above). If the cycles

(ijk), (123) have no symbols in common, then by two applications of the preceding argument we deduce that they are both conjugate to $(12k)$ (for instance) and so also to each other.

We next consider $\mathbf{A}_6$. Let H be a nontrivial normal subgroup of $\mathbf{A}_6$. We consider two cases.

(v) Suppose that no permutation in H other than 1 fixes any symbols (i.e. that the stabilizer in H of each symbol is trivial). Obviously there are then just two types of cycle decompositions possible: (12)(3456) and (123)(456). Permutations of the first type cannot exist in H since their squares are nontrivial and fix two symbols. It is easy to calculate that there are 40 permutations of the second type. Since in $\mathbf{A}_6$ they form either a single conjugacy class or else two conjugacy classes of the same size, we have that $|H| = 1 + 20r$, where $r = 1$ or 2. Hence $|H|$ does not divide $|\mathbf{A}_6|$, which is contrary to Lagrange's theorem; thus Case (v) is impossible.

(vi) Suppose that some nontrivial element x of H fixes a symbol i. Let A be the stabilizer of i in $\mathbf{A}_6$, i.e. the group of all elements of $\mathbf{A}_6$ fixing i. Clearly $A \simeq \mathbf{A}_5$, so that A is simple. Since $A \cap H$ is normal in A and nontrivial (since $x \in A \cap H$), we get that $A \cap H = A$. Hence $H \geq A$, so H contains a 3-cycle and therefore (by (iv)) coincides with $\mathbf{A}_6$.

Finally we consider $\mathbf{A}_n$ for $n \geq 7$. Again let H be a nontrivial normal subgroup of $\mathbf{A}_n$. We shall prove that $H = \mathbf{A}_n$. Let $1 \neq x \in H$, and let y be a 3-cycle which does not commute with x. Then $z = y^{-1}(x^{-1}yx)$ is a nontrivial element of H expressible as a product of two 3-cycles, and therefore moving at most 6 symbols. Let A be the subgroup consisting of all permutations in $\mathbf{A}_n$ fixing all but a certain 6 symbols, where these 6 symbols include those moved by z. Then $A \simeq \mathbf{A}_6$, and the proof may be completed as in Case (vi), using the simplicity of $\mathbf{A}_6$.

12.1.2. Exercise. Let $\mathfrak{n}$ be any cardinal number. The *alternating group of degree* $\mathfrak{n}$, written $\mathbf{A}_\mathfrak{n}$, consists of those permutations of a set of cardinal $\mathfrak{n}$ which move only finitely many symbols and can be written as a product of an even number of transpositions. By imitating the proof of the theorem of Galois, prove that $\mathbf{A}_\mathfrak{n}$ is simple provided that $\mathfrak{n} \neq 4$.

In connexion with this exercise note the following. Let M be a set of cardinal $\aleph_\nu$, and let $\mathbf{S}_\mu(M)$ be the set of all permutations in $\mathbf{S}(M)$ moving strictly fewer than $\aleph_\mu$ symbols. It is clear that the members of the chain

$$1 < \mathbf{A}(\aleph_\nu) < \mathbf{S}_0(M) < \mathbf{S}_1(M) < \cdots < \mathbf{S}_\nu(M) < \mathbf{S}(M), \qquad (1)$$

are all normal subgroups of $\mathbf{S}(M)$. It turns out that the members of the chain (1) are the *only* normal subgroups of $\mathbf{S}(M)$ (see [33]). In particular the quotient $\mathbf{S}(M)/\mathbf{S}_\nu(M)$ is simple.

12.2. The Projective Special Linear Groups

Let K be a field. The quotient group of $\mathbf{SL}_n(K)$ by its center (i.e. by the scalar matrices) is called the *projective special linear group* (of degree n over K), and is denoted by $\mathbf{PSL}_n(K)$. These groups—over finite fields—were added to the known series of groups by Jordan (1870), who established their simplicity. A certain lack of rigor in Jordan's original proof was later removed by Dickson.

12.2.1. Exercise. The linear fractional transformations

$$f(x) = \frac{ax+b}{cx+d}$$

with coefficients from the field K and determinant $ad - bc = 1$, form a group under composition of maps. This group is isomorphic to $\mathbf{PSL}_2(K)$.

12.2.2. Exercise. Verify the formula:

$$|\mathbf{SL}_n(q)| = \frac{1}{q-1} \prod_{i=0}^{n-1} (q^n - q^i); \tag{2}$$

$$|\mathbf{PSL}_n(q)| = \frac{1}{d(q-1)} \prod_{i=0}^{n-1} (q^n - q^i), \quad \text{where } d = (n, q-1). \tag{3}$$

(Hint. Consider $\mathbf{SL}$ as the kernel of the homomorphism det and use the formula (4) of the preceding section, and Exercise 3.1.6.)

We assume the reader is familiar with the basic facts about finite fields usually taught in general courses in algebra. We shall need in addition the following fact.

12.2.3. Lemma. *In any finite field K the equation $x^2 - y^2 = a$ is soluble for every a in K.*

PROOF. If the field K has characteristic 2, then its elements are just the roots of the equation $z^{2^m} - z = 0$, for some $m > 0$ (i.e. K is the splitting field of this polynomial over $\mathbf{Z}_2$). Hence every a in K is a square. If the characteristic is not 2, then the equations $x + y = a$, $x - y = 1$ have a simultaneous solution in K. That solution will then also be a solution of the equation $x^2 - y^2 = a$. (Note that for characteristic $\neq 2$ the argument works also for infinite fields K.)

12.2.4. Theorem (Jordan–Dickson). *For every finite field K the group $\mathbf{PSL}_n(K)$ is simple, with the exception of $\mathbf{PSL}_2(2)$ and $\mathbf{PSL}_2(3)$.*

PROOF. (i) Consider first the case $n = 2$. By (3) the groups $\mathbf{PSL}_2(2)$, $\mathbf{PSL}_2(3)$ have orders 6, 12, and are therefore not simple—see Exercise 11.2.1, and Exercise 13.2.4 (following this section).

Suppose that $|K| \geq 4$. Let H be a normal subgroup of $\mathbf{SL}_2(K)$ containing all the scalar matrices (that is, $\pm e$) and at least one non-scalar matrix a. We wish to prove that $H = \mathbf{SL}_2(K)$.

(ii) We show first that H contains a transvection of the form $t_{12}(\lambda)$ ($\lambda \neq 0$). Thus suppose first that $a_{21} = 0$, so that

$$a = \begin{pmatrix} \dfrac{1}{\alpha} & * \\ 0 & \alpha \end{pmatrix}.$$

If $\alpha^2 = 1$ then for the desired transvection we may take one of a or $-a$. If $\alpha^2 \neq 1$ then the commutator $[a, t_{12}(1)] = t_{12}(1 - \alpha^2)$ will serve. Suppose next that $a_{21} \neq 0$. For each nonzero $\beta \in K$ there is a unique field element $*$ such that the matrix

$$b = \begin{pmatrix} a_{11}\beta & * \\ a_{21}\beta & -a_{11}\beta \end{pmatrix}$$

lies in $\mathbf{SL}_2(K)$. Therefore H contains the matrix

$$c = -b^{-1}aba = \begin{pmatrix} \dfrac{1}{\beta^2} & * \\ 0 & \beta^2 \end{pmatrix}, \quad \text{for some } * \in K.$$

If $|K| \neq 5$ there exists β in K satisfying $\beta^4 \neq 1$, so that for the desired transvection we may again take the commutator $[c, t_{12}(1)] = t_{12}(1 - \beta^4)$. It remains to analyse the case $|K| = 5$. Set $\beta = 1$; then $c = t_{12}((2/a_{21})(a_{11} + a_{22}))$. If the trace of $a = a_{11} + a_{22} \neq 0$, then c will serve. If $\operatorname{tr} a = 0$, then take in place of a the matrix $a^* = [a, t_{12}(1)]$, for which we have

$$\operatorname{tr} a^* = 2 + a_{21}^2 \neq 0.$$

(iii) The next step is to show that H contains all transvections and so coincides with $\mathbf{SL}_2(K)$. In the first place H contains all transvections of the form $t_{12}(*)$ since

$$\begin{pmatrix} \dfrac{1}{\rho} & 0 \\ 0 & \rho \end{pmatrix}^{-1} t_{12}(\lambda) \begin{pmatrix} \dfrac{1}{\rho} & 0 \\ 0 & \rho \end{pmatrix} = t_{12}(\lambda\rho^2),$$

$$t_{12}(\lambda\rho^2) t_{12}(\lambda\sigma^2)^{-1} = t_{12}(\lambda(\rho^2 - \sigma^2)),$$

and $\rho^2 - \sigma^2$ takes all possible values as ρ, σ vary over K (see 12.2.3). Finally

$$\begin{pmatrix} 0 & -1 \\ 1 & 0 \end{pmatrix}^{-1} t_{12}(\lambda) \begin{pmatrix} 0 & -1 \\ 1 & 0 \end{pmatrix} = t_{21}(-\lambda).$$

This completes the proof for the case $n = 2$.

(iv) We now consider the case $n \geq 3$ for *any* field K. Again let H be a normal subgroup of $\mathbf{SL}_n(K)$ containing all scalar matrices and at least one nonscalar matrix a; we have to prove that $H = \mathbf{SL}_n(K)$. (The reader may find it useful for understanding the following argument to write each matrix w in the form $w = \sum w_{rs} e_{rs}$ and use the multiplication rule (1) of 3.1.1.(IV).) Since a is nonscalar there is at least one transvection $t_{ij}(1)$ with which it does not commute. By re-indexing the rows and columns if necessary, we may assume that $x = [a, t_{21}(1)] \neq e$. If we write $y = a^{-1}$, then by matrix multiplication we get

$$x = e - \begin{pmatrix} * & y_{12}a_{12} & \cdots & y_{12}a_{1n} \\ * & y_{22}a_{12} & \cdots & y_{22}a_{1n} \\ \cdots\cdots\cdots\cdots\cdots\cdots\cdots\cdots\cdots \\ * & y_{n2}a_{12} & \cdots & y_{n2}a_{1n} \end{pmatrix}.$$

(v) We shall deduce from this that H contains a nonidentity matrix z of the form

$$z = \left(\begin{array}{cc|c} * & * & \\ * & * & 0 \\ \hline & * & e \end{array} \right).$$

Thus if $a_{13} = \cdots = a_{1n} = 0$, we may take $z = x$. If on the other hand some $a_{1k} \neq 0$, $k \neq 1, 2$, then take $z = f_{2k}^{-1} u^{-1} x u f_{2k}$, where

$$u = e - \frac{1}{a_{1k}} \sum_{\substack{i=2 \\ i \neq k}}^{n} a_{1i} e_{ki}; \qquad f_{2k} = e - e_{22} - e_{kk} + e_{2k} - e_{k2}.$$

(vi) From the existence in H of such a matrix z we can deduce in turn that there is in H a nonidentity matrix

$$v = \left(\begin{array}{ccc|c} 1 & 0 & 0 & \\ 0 & 1 & 0 & 0 \\ * & * & 1 & \\ \hline & 0 & & e \end{array} \right).$$

To see this note first that

$$[z, t_{31}(1)] = e - (z_{11} - 1)e_{31} - z_{12}e_{32},$$
$$[z, t_{32}(1)] = e - z_{31}e_{31} - (z_{22} - 1)e_{32}.$$

If either of these commutators is different from e, it will serve as v. Suppose they are both e. Then

$$z = \left(\begin{array}{cc|c} 1 & 0 & \\ 0 & 1 & 0 \\ \hline & * & e \end{array} \right).$$

If $n = 3$ or $z_{i1} = z_{i2} = 0$ for all $i \geq 4$, then we may take $v = z$. If for instance both $z_{41}, z_{42} \neq 0$, then for v we may take the commutator

$$[z, t_{34}(1)] = e - z_{41}e_{31} - z_{42}e_{32}.$$

(vii) The next step is to show that H contains a transvection. If $v_{32} = 0$ then v is a transvection. If $v_{32} \neq 0$ then we may take

$$t_{21}(\lambda)^{-1}vt_{21}(\lambda) = e + (v_{31} + \lambda v_{32})e_{31} + v_{32}e_{32},$$

with the obvious λ.

(viii) Finally we show that H contains all transvections and therefore coincides with $\mathbf{SL}_n(K)$. This is immediate from the equation

$$f_{ik}^{-1}t_{ij}(\lambda)f_{ik} = t_{kj}(\lambda), \qquad f_{lj}^{-1}t_{kj}(\lambda)f_{lj} = t_{kl}(-\lambda),$$

$$d_{ik}(\mu)^{-1}t_{ij}(\lambda)d_{ik}(\mu) = t_{ij}(\lambda\mu),$$

where i, j, k, l are pairwise distinct symbols, $f_{ik} = e - e_{ii} - e_{kk} + e_{ik} - e_{ki}$, and $d_{ik}(\mu)$ is the diagonal matrix with $(1/\mu)$ as the ith diagonal entry, μ as the kth diagonal entry, and the remaining diagonal entries all 1. This completes the proof of the theorem.

The theorem is in fact true without the restriction that the field be finite: we have actually proved this stronger result for $n \geq 3$, since in Parts (iv) to (viii) of the proof the finiteness of the field was not used. For $n = 2$ our proof depended on a lemma (12.2.3) where the finiteness of the field is essential— however there are ways of avoiding the use of this lemma (see [13], [32]).

In connexion with the methodology of proof we make a further remark. In §12.1 we managed to establish the simplicity of $\mathbf{A}_5$ in a rather heavy-handed way: we enumerated each of its conjugacy classes and discovered that the only subsets of $\mathbf{A}_5$ which are unions of conjugacy classes, and have orders dividing the order of $\mathbf{A}_5$, are the whole group and the identity subgroup. The chain of argument from (i) to (iv) above is more typical, and in principle more universally applicable, as a method for proving simplicity: In our normal subgroup H we chose an element a about which we knew only that it was not the identity, and then, using the closure properties of H under multiplication, forming inverses, conjugation by any element, and the derivative operation of forming commutators with any element, by degrees we brought to light more and more elements of H until finally we discovered in H generators for the whole group. If a philosopher were to stoop to the contemplation of the Jordan–Dickson theorem, he might summarize the idea of the proof as follows: Getting hold of one link, we pull out the whole chain. Unfortunately this guiding principle leaves obscure exactly how the chain can be pulled out.

12.2.5. Exercise. In view of the Jordan–Dickson theorem it is natural to ask whether we obtain new simple finite groups if in the definition of $\mathbf{PSL}_n(K)$

we take K to be the ring $\mathbf{Z}_m$ rather than a field. This turns out to be a vain hope: of course if m is prime then $\mathbf{Z}_m$ is a field, but if m is composite then the group $\mathbf{PSL}_n(\mathbf{Z}_m)$ is not simple.

12.2.6. Exercise. Find a group whose central quotient is simple, but whose center is not a direct factor. (Solution. If in the group $G = \mathbf{SL}_2(5)$, the center C, consisting of the matrices $\pm e$, had a direct complement H, then the square of every element of G would lie in H. In particular the square of the matrix

$$\begin{pmatrix} \dfrac{1}{\zeta} & 0 \\ 0 & \zeta \end{pmatrix}$$

where ζ is a generator of the multiplicative group of the field of 5 elements, would lie in H. However this square is just $-e$, so that we would have that $-e \in C \cap H$, which is impossible.)

J. G. Thompson [2-signalizers of finite groups, Pacific J. Math. **14** (1964), 363–364] coined the term 2-*signalizer* of a finite group G for a subgroup A of G with the property that both $|A|$ and $|G : N_G(A)|$ are odd. In the same paper he conjectured that the 2-signalizers of simple groups are always abelian. We conclude this section with a counterexample to this conjecture due to V. D. Mazurov [On 2-signalizers of finite groups, Algebra i Logika **7** (1968), 60–62]. It uses only the simplicity of $\mathbf{PSL}$.

12.2.7. EXAMPLE. The simple group $\mathbf{PSL}_7(q)$, q odd, contains nonabelian 2-signalizers. To see this consider the subgroups A, B of $G = \mathbf{SL}_7(q)$ consisting respectively of all matrices of the forms

$$\begin{pmatrix} 1 & * & * \\ 0 & e_2 & * \\ 0 & 0 & e_4 \end{pmatrix}, \quad \begin{pmatrix} \alpha(x, y) & 0 & 0 \\ 0 & x & 0 \\ 0 & 0 & y \end{pmatrix}, \quad \alpha(x, y) = \frac{1}{\det x \cdot \det y},$$

where e_k is the identity of degree k, and the two matrices are decomposed into blocks in the same way. In view of the formula (2) of this section and (4) of the preceding, we have

$$|G| = \frac{1}{q-1}(q^7 - 1)(q^7 - q) \cdots (q^7 - q^6),$$

$$|A| = \text{a power of } q,$$

$$|B| = (q^2 - 1)(q^2 - q)(q^4 - 1)(q^4 - q)(q^4 - q^2)(q^4 - q^3).$$

It is immediate that $|A|$ and $|G : B|$ are odd. It is also clear that B normalizes A, so that $|G : N_G(A)|$ is odd. If ϕ is the natural homomorphism from G to G/C where C is the center of G, then of course the order of A^ϕ and the index of its normalizer in G^ϕ will still be odd, i.e. A^ϕ is a 2-signalizer in $\mathbf{PSL}_7(q)$. Finally, since $A \cap C = 1$, we have that $A^\phi \approx A$, so that A^ϕ is nonabelian.

§13. Permutation Groups

To conclude the chapter we shall look at groups of permutations. These were the historical prototypes for group theory, and they are still important. They were first introduced by Évariste Galois in connexion with his investigation of the conditions under which algebraic equations are soluble by radicals: With each algebraic equation he associated a certain group of permutations of the roots, from certain of whose properties the question of the solubility in radicals of the original equation can be answered. In 1870 there appeared Jordan's fundamental tract on permutation groups which contained a clear and detailed exposition of Galois' ideas, and also many results on permutations. Only towards the beginning of this century did the modern abstract group concept shed these swaddling clothes, and from then on the theory of permutation groups gradually assumed its more modest position in the general theory. Generally speaking permutation groups arise naturally whenever the symmetries of objects of "finite type" are studied—for example with each crystal we may associate its group of symmetries, described by permutations of the crystal's vertices. The reader wishing to study permutation groups more deeply may consult the book [43].

13.1. The Regular Representation

It turns out that permutation groups exhaust the stock of finite groups.

13.1.1. Theorem (Cayley). *Every finite group is isomorphic to some group of permutations.*

PROOF. Let G be a finite group that we wish to represent faithfully by permutations. Permutations of what?—This is the first question that comes to mind. Since there is nothing else on hand let us try G itself as the set whose elements are to be permuted. Let g be an element of G. What permutation π_g shall we associate with g? If we write a permutation in the normal way as two rows of elements with each element of the bottom row the image of the element above it, then we have the top row of π_g—it consists of the elements of G in some order. Multiply then on the right by g one at a time and write the results in the second row. It follows from the group axioms that the elements of the bottom row are pairwise distinct, so that π_g is indeed a permutation. It remains to check that the map defined by $g \to \pi_g$ is a monomorphism. In the first place it is a homomorphism, i.e. $\pi_{ab} = \pi_a \pi_b$, since on each $x \in G$ both sides of the equality act in the same way:

$$x\pi_{ab} = x(ab) = (xa)b = (xa)\pi_b = (x\pi_a)\pi_b = x(\pi_a \pi_b).$$

Secondly, the kernel is trivial: if π_g is the identity map then on the one hand it fixes 1, and on the other maps it to g. Hence $g = 1$, and the theorem is proved.

13.1.2. Exercise. Every finite group can be embedded in a finite simple group.

The reader may have noticed that in the proof of Cayley's Theorem the finiteness of G is not used. Thus *every* group G is faithfully represented by its right translations—this is called *the regular representation* of the group G.

13.1.3. Generalization of Cayley's Theorem. *Every group is isomorphic to a group of one-to-one maps ("infinite permutations") of some set onto itself.*

The representation of abstract groups by permutations is also useful in a wide variety of questions outside of finite group theory.

13.1.4. EXAMPLE. *Embedding theorems.* It is often necessary to embed a given group G into a bigger group G^* with one or another interesting property—for example one of the following properties:

(i) that G^* is simple;

(ii) that in G^* extraction of roots of any degree is possible; i.e. that the equation $x^n = g$ is soluble in G for all $g \in G$ and every integer $n > 0$ (cf. divisibility in abelian groups, §9).

(iii) that every pair of elements of G^* with the same order (finite or infinite) are conjugate in G^*.

The reader, guided by his own idea of which properties are interesting, will easily be able to add to this list.

In two papers by the authors and V. N. Remeslennikov [On constructing closures of groups, Dokl. Akad. Nauk SSSR **134**, No. 3 (1960), 518–520; On a method of constructing closures of groups, Uč. zap. Permskogo un-ta **17**, No. 2 (1960), 9–11] a method for proving embedding theorems was set forth. The method consists in embedding a group supposed given as a group of permutations in a larger permutation group possessing *to a greater degree* the desired property (simplicity, roots, conjugating elements etc.), after which the construction is completed in the well-known manner by inductively iterating the embedding, if necessary infinitely often. The virtue of this method lies in the fact that the concrete representation of a group by means of permutations makes computations more concrete and therefore easier. To illustrate this we shall show how to embed an arbitrary group G in a group G^* with property (iii). We define a chain of groups

$$G = G_0 \leq G_1 \leq G_2 \leq \cdots,$$

inductively by taking G_{n+1} to be the group of all permutations of G_n, i.e. the symmetric group on the set G_n, and identifying G_n with its regular representation. The desired group G^* is then just the union (or direct limit) of the G_n.

To see this let a, b be two elements of the same order r (which may be infinite) in G^*; they lie in some G_n and thus as elements of G_{n+1} they are each products of the same number of disjoint cycles of length r. By 2.5.7 they are therefore conjugate in G_{n+1} and hence in G^*.

Similarly if we wish to embed an arbitrary group G in a group G^* with property (ii), then it suffices to embed G in a group in which the equation $x^n = g$ is soluble for any single $g \in G$ and any single n. One way to achieve this is given, in the language of permutations, in the aforementioned papers.

Finally we sketch the solution to (i). The case that the group G is finite is just Exercise 13.1.2 above. Suppose G is infinite, say $|G| = \aleph_\nu$, and let $\mathbf{S}_\nu(G)$ be the group of permutations of G moving fewer than $|G|$ elements (as in (1) of the preceding section). The group G identified with its regular representation is contained in $\mathbf{S}(G)$ and intersects $\mathbf{S}_\nu(G)$ trivially, and therefore can be embedded in the simple group $\mathbf{S}(G)/\mathbf{S}_\nu(G)$.

It is natural to ask if an arbitrary group G can be embedded in a group G^* having at once properties (i), (ii), etc. The answer is an easy affirmative provided that for each property separately there is a group with that property in which G can be embedded, and if each property is preserved in unions of increasing chains of groups with the property: for then all one needs to do is to iterate the embeddings of types (i), (ii), (iii), . . . a countable infinity of times, and take the union.

Other methods of obtaining results about embeddings—in particular embeddings into groups with properties (i), (ii) and (iii)—have been given by: P. Hall [Some constructions for locally finite groups, J. London Math. Soc. **34** (1959), 305–319], G. Higman, B. H. Neumann and Hanna Neumann [Embedding theorems for groups, Ibid. **24** (1949), 247–254] and B. H. Neumann [Adjunction of elements to groups, Ibid. **18** (1943), 4–11]; Hall's paper uses wreath products, the second paper introduces a celebrated construction now called an "HNN extension," and B. H. Neumann's paper involves the related "free product with an amalgamated subgroup."

13.2. Representations by Permutations of Cosets

We shall now generalize Cayley's Theorem in a different direction. Let H be a subgroup of finite index in a group G, itself not necessarily finite. With each element g of G we associate the permutation π_g of the set of right cosets of H in G, defined, analogously to §13.1, by

$$\pi_g = \begin{pmatrix} Hx_1 & \cdots & Hx_m \\ Hx_1g & \cdots & Hx_mg \end{pmatrix},$$

where $\{x_1, \ldots, x_m\}$ is a complete set of right coset representatives for H in G.

13.2.1. Theorem. *The mapping $\pi: g \to \pi_g$ defined above is a homomorphism from G to $\mathbf{S}(G/H)$. The kernel is the largest normal subgroup of G contained in H.*

PROOF. It is easily verified directly, as before, that π is a homomorphism. Let K denote the kernel of this homomorphism, and N the largest normal subgroup of G contained in H (sometimes called the "core" of H in G). We show first that $K \le N$. Let $g \in K$, so that $\pi_g = 1$. This means that $Hxg = Hx$ for all $x \in G$. From this we get that $H^x g = H^x$, whence

$$g = \bigcap_{x \in G} H^x.$$

It is clear that this intersection is a normal subgroup of G contained in H. It is equally clear that a normal subgroup of G contained in H must be contained in every H^x. Hence

$$N = \bigcap_{x \in G} H^x.$$

Thus $g \in N$ and $K \le N$. For the reverse inclusion, let $g \in N$; then $Hxg = Hxgx^{-1}x = Hx$, so that $\pi_g = 1$ and $N \le K$, completing the proof.

From this and Lagrange's theorem (2.4.5) we obtain immediately

13.2.2. Theorem (Poincaré). *Every subgroup of finite index m of a group G contains a normal subgroup of G of finite index divisible by m and dividing $m!$.*

A weaker result than this was obtained in Chapter 1 (see 2.5.13). Theorem 13.2.1, and its consequence Poincaré's theorem, are useful for proving that in a given group there are certain divisors of the group order which none the less are excluded from being the orders of subgroups.

13.2.3. Exercise. The alternating group $\mathbf{A}_5$ of order 60 has no subgroups of order 15, 20 or 30.

(Solution. If $\mathbf{A}_4$ had a subgroup of order 15, that is of index 4, then by Poincaré's theorem it would have to contain also a normal subgroup of index dividing 24 and divisible by 4. This however is not possible since by the theorem of Galois $\mathbf{A}_5$ is simple. The other cases are treated similarly.)

13.2.4. Exercise. No group of order 12 can be simple. In particular therefore the group $\mathbf{PSL}_2(3)$ is not simple—this was mentioned in the preceding section.

Let R be a ring and H a group. The *group ring* of the group H over the ring R is the ring $R[H]$ of all formal expressions

$$r_1 h_1 + \cdots + r_n h_n, \qquad r_i \in R, \quad h_i \in H,$$

(where $rh = 0$ if $r = 0$), with addition and multiplication given by:

$$\sum r_i h_i + \sum s_i h_i = \sum (r_i + s_i) h_i,$$

$$\sum r_i h_i \cdot \sum s_j f_j = \sum (r_i s_j)(h_i f_j).$$

(We met with a special case of this concept in Example 3.2.6. We shall need this concept in the following discussion.)

To return to our original theme (with H as a finite index subgroup of G), notice that the representation of G given by $g \to \pi_g$, associates with each g of G a permutation $\pi(g)$ of the indices $1, 2, \ldots, m$, of the coset representatives, and a set of "factors" or "multipliers" $h_i(g) \in H$, given by

$$Hx_i g = Hx_{i\pi(g)}, \qquad x_i g = h_i(g) x_{i\pi(g)}.$$

If we write $\hat{\pi}(g)$ for the $m \times m$ "permutation matrix" corresponding to the permutation $\pi(g)$, with entries 0 or 1, then it is straightforward to check that

$$g \to \mathrm{Diag}(h_1(g), \ldots, h_m(g)) \cdot \hat{\pi}(g)$$

defines a monomorphism $G \to \mathbf{GL}_m(\mathbf{Z}[H])$, where $\mathbf{Z}[H]$ is the group ring of H over the integers. This embedding is called the *monomial representation of the group G over the subgroup H*. We see from this that every group having H (or a group isomorphic to H) as a subgroup of index m is embeddable in a group of all those matrices of degree m over $\mathbf{Z}[H]$ having in each row and column exactly one nonzero entry, and that from H (the *monomial subgroup* of $\mathbf{GL}_m(\mathbf{Z}[H])$. This embedding, due to the German mathematician G. Frobenius (active around 1900), is much used in the (linear) representation theory of finite groups, where it yields the representation of a group "induced" by a representation of a subgroup.

All this can be translated into the language of wreath products as follows.

Let A, B be groups with $B \le S(X)$ for some set X. It is easy to verify that the set $B \times A^{[X]}$ under the multiplication

$$bf \cdot b_1 f_1 = bb_1 f^{b_1} f_1, \quad \text{where } f^b(x) = f(xb^{-1}), \ x \in X,$$

is a group. It is called the *wreath product of the group A with the permutation group B*. The most important case of this is the (unrestricted) wreath product which we met with in Chapter 2; in the present context this is called the *standard* wreath product, and is obtained by taking X to be B itself and letting B act on itself according to the rule $x \to b^{-1}x$. Using for the more general "permutational" wreath product the same notation $A \,\mathrm{Wr}\, B$ as previously, with the understanding that here B acts on a set X, we see that the monomial subgroup of $\mathbf{GL}_m(\mathbf{Z}[H])$ is isomorphic to $H \,\mathrm{Wr}\, S_m$, and that, at least when the active group is finite, Theorem 6.2.8 is just the special case of the above embedding when this wreath product is standard.

13.3. Transitivity. Primitivity

Here we shall consider some concepts having to do with the very essence of the action of a permutation group on the set of symbols it permutes. In fact these concepts make sense for the more general situation of a group acting on a set (see §11.1 for the definition), so we shall give the definitions in that more general context.

If a group G acting on a set M has just one orbit—M itself—then we say that G acts *transitively* on M. It is clear that this is the same as requiring that for any two elements, m, m' of M there is an element g of G such that $mg = m'$. Generalizing this, we say that the group G is *r-fold transitive* if for any two ordered r-tuples $(m_1, \ldots, m_r)$ and $(m'_1, \ldots, m'_r)$ of elements of M, where $m_i \neq m_j$ and $m'_i \neq m'_j$ for $i \neq j$, there exists an element g in G such that $m_i g = m'_i$, $1 \leq i \leq r$.

A partition $\{M_\alpha\}$ of the set M into (pairwise disjoint) subsets M_α, is called a *partition into blocks relative to* G, if for each M_α and each $g \in G$ there is an M_β in the partition such that $M_\alpha g = M_\beta$, i.e. $M_\alpha g$ is again in the partition, or in other words the blocks M_α are permuted as wholes by the elements of G. Of course there are always trivial partitions of M into blocks: the partition of M into one-element subsets, and the partition into just the single block M. If there are no nontrivial partitions of M into blocks then we say that the group G is *primitive*.

Of course these concepts apply in particular to ordinary permutation groups permuting a finite set M.

13.3.1. Exercise. In the definition of blocks the condition that $M_\alpha g = M_\beta$ may be weakened to $M_\alpha g \subseteq M_\beta$.

13.3.2. Exercise. The groups $\mathbf{S}_n$ and $\mathbf{A}_n$ are respectively n-fold and $(n-2)$-fold transitive.

13.3.3. Exercise. Every twofold transitive permutation group is primitive.

13.3.4. Exercise. Every nontrivial normal subgroup N of a primitive permutation group G, is transitive. (Hint. The orbits of N constitute a partition into blocks for G.)

Suppose that groups G, $\hat{G}$ act on sets M, $\hat{M}$ respectively. It is natural to call these groups (with their actions implicit) *isomorphic*—as groups acting on sets—if there is a one-to-one map from M onto $\hat{M}$ (say ϕ), and an isomorphism between G and $\hat{G}$ (say ψ), such that corresponding elements of the groups send corresponding elements of the sets acted on, again to corresponding elements, i.e.

$$m\phi g^\psi = (mg)\phi \quad \text{for all } m \in M, g \in G.$$

Permutation groups isomorphic in this sense are sometimes also called *similar*.

In the preceding section (§13.2) we showed how each subgroup of finite index gives rise to a representation of the group by permutations of the set of right cosets. It is easy to see that this representation is transitive. As it turns out the converse of this is true: Every transitive permutation representation of a given group is similar to the representation of the group by permutations of the right cosets of some subgroup of finite index, as described in §13.2. This is the gist of the following theorem.

13.3.5. Theorem. *Let π_1 be an epimorphism of a group G onto a transitive group of permutations of a finite set M. Define an action of G on M in the natural way by the rule $mg = mg^{\pi_1}$. Let $m_1 \in M$ and denote by H the stabilizer of m_1 in G, i.e.*

$$H = \{g \mid g \in G, m_1 g = m_1\}.$$

Then H is a subgroup of finite index in G, and the permutation group G^{π_1} is similar to the group of permutations of the right cosets of H in G, described in §13.2.

PROOF. Write $M = \{m_1, \ldots, m_s\}$, and

$$H_i = \{g \mid g \in G, m_1 g = m_i\}, \qquad 1 \le i \le s.$$

Since G^{π_1} is transitive the H_i are all nonempty. It can be verified immediately that H_1 is a subgroup of G and that the H_i are just the right cosets of H_1 in G. Since $H = H_1$ we obtain that H is a subgroup of finite index in G.

Let π be, as in §13.2, the representation of G by permutations of the right cosets of its subgroup H. It remains to show that the permutations G^{π_1} and G^π are similar. Let ϕ be the map from M onto the set of right cosets of H in G which sends m_i to H_i, and let ψ be the map of G^{π_1} onto G^π sending g^{π_1} to g^π. It is easily seen that if $x^{\pi_1} = y^{\pi_1}$, then $x^\pi = y^\pi$, so that the map ψ is well-defined. It is then immediate that ϕ, ψ are one-to-one and onto, and that

$$(m_i\phi)g^\pi = (m_i g^{\pi_1})\phi \quad \text{for } 1 \le i \le s, g \in G,$$

completing the proof.

13.3.6. Exercise. State a form of Theorem 13.3.5 without the condition that M be finite.

We shall now show that the study of arbitrary permutation groups reduces in some sense to the study of transitive groups, and that this in turn reduces to the study of primitive groups. Recall that a subdirect product of groups G_i is a subgroup of the direct product of the G_i, whose projection on each factor G_i is the whole of G_i.

13.3.7. Theorem. (*i*) *Every group of permutations of n symbols is a subdirect product of transitive groups of permutations of n_i symbols, where $\sum n_i = n$.*

(*ii*) *Let G be a transitive group of permutations of a set M, and let M' ⊆ M be a block from a partition of M into minimal blocks with at least* 2 *elements. Then the stabilizer of M', i.e. the set*

$$H = \{g \mid g \in G, M'g = M'\},$$

is a subgroup of G which acts primitively on M'.

PROOF. (i) Let G be a group of permutations of a set M, with orbits M_i, and for each i let G_i be the group of permutations of M_i obtained by restricting G to M_i. It is easy to see that G is a subdirect product of the G_i.

(ii) Clearly H is a subgroup of G. Let $\{x_1, \ldots, x_s\}$ be a complete set of right coset representatives for H in G. Then the blocks of the given partition are just $M'x_1, M'x_2, \ldots, M'x_s$ (here the transitivity is being used). If the restriction of H to M' were not primitive then we could partition M' into smaller blocks $M'_1, M'_2, \ldots, M'_r$, relative to H; but then our original partition of M into G-blocks could be refined to the partition $M'_i x_j$, $1 \le i \le r$, $1 \le j \le s$, contradicting the minimality of the original partition.

13.3.8. Theorem. *Let G be a transitive group of permutations of a set M, let M_1 be a block in a partition of M into blocks, and let $m_1 \in M_1$. Denote by H the stabilizer of m_1 and by K the stabilizer of M_1, i.e.*

$$H = \{g \mid g \in G, m_1 g = m_1\},$$

$$K = \{g \mid g \in G, M_1 g = M_1\}.$$

Then H and K are subgroups of G with $H \le K \le G$, the number of blocks in the given partition is $|G:K|$, and each block contains $|K:H|$ elements.

PROOF. That $H \le K \le G$ is immediate from the definitions. For the rest let $Hx_1, \ldots, Hx_s$ be the right cosets of H in G, with $x_1 = e$, the identity. By Theorem 13.3.5 we may take M to be $\{Hx_1, \ldots, Hx_s\}$ with G acting by multiplication on the right. Suppose that $M_1 = \{Hx_1, \ldots, Hx_r\}$, and $m_1 = Hx_1 = H$ (by choosing M_1 to be the block containing H and reindexing the x_i's if necessary). Since G is transitive each G-block is obtained from M_1 by right multiplication by an element of G; thus all blocks have the same size, and it remains to show only that $r = |K:H|$, or that

$$K = Hx_1 \cup Hx_2 \cup \cdots \cup Hx_r.$$

Now on the one hand any element of the form hx_i, $h \in H$, $1 \le i \le r$, sends the coset H to some coset in M_1, and therefore sends the whole block M_1 into itself. Hence

$$Hx_1 \cup \cdots \cup Hx_r \subseteq K.$$

On the other hand if $x \in K$ then $Hx = Hx_i$ for some $1 \le i \le r$, whence the reverse inclusion.

13.3.9. Exercise. A transitive permutation group G is primitive if and only if the stabilizer of some letter (or, equivalently, each letter) is a maximal subgroup of G.

5 Free Groups and Varieties

§14. Free Groups

At the beginning of Chapter 3 we introduced the concept of a group free in a given class of groups, and showed that in the class of all abelian groups free groups exist and can be explicitly described. We shall in this section prove the existence of free groups in the class of *all* groups (also called "absolutely free" or simply "free" groups), and also give an internal description of them and investigate their simplest properties.

14.1. Definition

Given a set of generators S of a group G, there are always relations between these generators; i.e. various products of the generators and their inverses will be equal to the identity element of G. Among such relations there will inevitably occur the following: $xx^{-1} = e$, $x^{-1}x = e$, where x ranges over S. Of course the reason why these always occur is that they are a consequence of the group axioms; we call these relations, and their like, *trivial*. Now it turns out that there are groups possessing sets of generators on which there are no relations other than the trivial ones—in other words these generators are "free of relations" (which, incidentally, explains the origin of the name "free"). Our immediate aim is to show how to construct such groups, and to prove that they are indeed free in the class of all groups, in the sense of the general definition given in Chapter 3.

For any group G with generators g_i, $i \in I$ (some index set), we think loosely of the elements of G as being "words" $g_{i_1}^{\varepsilon_{i_1}} \cdots g_{i_m}^{\varepsilon_{i_m}}$, $\varepsilon_j = \pm 1$, and of multiplication as juxtaposition of words. This hints at the idea behind the construction: we shall declare the elements of the free group to be words in the letters (or symbols) x_i, x_i^{-1}, $i \in I$, which do not have segments (i.e. subwords) of the form $x_i^\varepsilon x_i^{-\varepsilon}$, $\varepsilon = \pm 1$, and their juxtaposition followed by deletion of the prohibited subwords, will define the operation.

We now give the precise definition. Write

$$X = \{x_i \mid i \in I\} \quad \text{and} \quad X^{-1} = \{x_i^{-1} \mid i \in I\}.$$

A *word* in the *alphabet* X is a finite sequence of symbols from $X \cup X^{-1}$. (In particular the empty word (or sequence) will be denoted by e.) The length of the sequence is called the *length* of the word. We say that a word is *reduced* if it does not contain adjacent symbols of the form x_i^ε, $x_i^{-\varepsilon}$, $\varepsilon = \pm 1$. For example the first of the words

$$x_2 x_1 x_1 x_2^{-1} x_3, \qquad x_1 x_2 x_2^{-1} x_3$$

is reduced, while the second is not reduced. Two words u, v will be called *equivalent* (in symbols: $u \sim v$), if v can be obtained from u by a finite number of insertions and deletions of words of the form $x_i^\varepsilon x_i^{-\varepsilon}$. It is clear that the relation $\sim$ is an equivalence on the set of all words; we shall denote by $[u]$ the equivalence class of all words equivalent to the word u.

14.1.1. Theorem. *Let* $X = \{x_i \mid i \in I\}$. *If on the set* $\mathbf{F}(X)$ *of classes of equivalent words in the alphabet* X *we define multiplication by* $[u][v] = [uv]$, *then this multiplication is well-defined (i.e. is independent of the choice of representatives of the classes). The set* $\mathbf{F}(X)$ *with this multiplication is a group.*

PROOF. (i) Each class of equivalent words contains a unique reduced word. To see this let $\rho(u)$ denote the reduced word obtained from u by successive deletions, moving from right to left, of subwords of the form $x_i^\varepsilon x_i^{-\varepsilon}$, The function ρ has the following properties (here $\equiv$ means "equal as written," "identical as words"; we wish to reserve the symbol $=$ for later use as meaning "equal as elements in the free group"):

$$\rho(u) \sim u, \tag{1}$$

$$\rho(u) \equiv u, \quad \text{if } u \text{ is reduced,} \tag{2}$$

$$\rho(uv) \equiv \rho(u\rho(v)), \tag{3}$$

$$\rho(x_i^\varepsilon x_i^{-\varepsilon} u) \equiv \rho(u) \quad \text{for } \varepsilon = \pm 1, \tag{4}$$

$$\rho(u x_i^\varepsilon x_i^{-\varepsilon} v) \equiv \rho(uv) \quad \text{for } \varepsilon = \pm 1, \tag{5}$$

$$\rho(uv) \equiv \rho(\rho(u)\rho(v)). \tag{6}$$

Properties (1), (2), (3) are immediate from the definition of ρ; (4) follows from (3); (5) follows from (3) and (4); and finally (6) follows from (3), (4), (5) by induction on the length of u. Suppose now that $u \sim v$, where u, v are reduced words. By definition there is a sequence

$$u \equiv u_1, u_2, \ldots, u_m \equiv v$$

of words, such that neighboring words differ by a single subword of the form $x_i^\varepsilon x_i^{-\varepsilon}$. Therefore by (5) we have that $\rho(u_i) \equiv \rho(u_{i+1})$, whence $\rho(u) \equiv \rho(v)$. Since u and v are reduced this implies that $u \equiv v$.

(ii) The fact that the product $[u][v]$ is independent of the choice of representatives u, v, is immediate from (i) and (6). The associativity of the operation is clear from the definition. The identity element is the equivalence class containing the empty word, and the inverse of the class $[x_{i_1}^{\varepsilon_1} \cdots x_{i_m}^{\varepsilon_m}]$ is the class $[x_{i_m}^{-\varepsilon_m} \cdots x_{i_1}^{-\varepsilon_1}]$. This completes the proof.

The group $\mathbf{F}(X)$ is called the *free group freely generated by the set X*, and the cardinal number $|X|$ the *rank* of the free group. We shall sometimes refer to X as a *free basis* for $\mathbf{F}(X)$. In the cases $X = \{x_1, \ldots, x_n\}$, $X = \{x_1, x_2, \ldots\}$, one often writes $\mathbf{F}_n(X)$, $\mathbf{F}_\infty(X)$ respectively, instead of $\mathbf{F}(X)$. If explicit indication of the free generators is unnecessary, one may write simply $\mathbf{F}_n$, $\mathbf{F}_\infty$. From now on we shall indulge in the usual abuse of language and use representatives to denote classes; thus we shall write for instance $u = v$, $uv = w$, instead of $[u] = [v]$, $[u][v] = [w]$. In view of (i), we may use the reduced words as representatives of the equivalence classes, and may thus talk of the "reduced form" of a word u, or the class containing u, meaning by this of course $\rho(u)$.

14.1.2. Exercise. Free groups of rank ≥ 2 are noncommutative.

14.1.3. Exercise. If u is not the empty word, then for $n \geq 2$ the length of the word $\rho(u^n)$ exceeds the lengths of the words $\rho(u)$, $\rho(u^{-1})$. Hence free groups are torsion-free.

14.1.4. Exercise. A reduced word is said to be *cyclically reduced* if the symbols it begins and ends with are not inverses of one another; i.e. if it begins with x_i^ε then it should not end in $x_i^{-\varepsilon}$. If we delete successively from the beginning and end of an arbitrary word u, the symbols violating this condition, then after a finite number of steps we shall arrive at the cyclically reduced form $\sigma(u)$ of the word u. Prove that elements u, v of a free group are conjugate if and only if there exist words w_1, w_2 such that $\sigma(u) \equiv w_1 w_2$, and $\sigma(v) \equiv w_2 w_1$.

It turns out that any group generated by a set of cardinal $\mathfrak{n}$, is an epimorphic image of the free group of rank $\mathfrak{n}$. More than that, the groups $\mathbf{F}(X)$ are precisely the groups free in the class of all groups, as the next theorem shows.

14.1.5. Theorem. *Suppose a group G is generated by a set $M = \{g_i \mid i \in I\}$, and let X be an alphabet $\{x_i \mid i \in I\}$. The map $X \to M$ defined by $x_i \to g_i$, extends to a unique epimorphism $\mathbf{F}(X) \to G$.*

PROOF. It is clear that as the image of the class $[x_{i_1}^{\varepsilon_1} \cdots x_{i_m}^{\varepsilon_m}]$ we shall have to take the element $g_{i_1}^{\varepsilon_1} \cdots g_{i_m}^{\varepsilon_m}$. That this does define a map from $\mathbf{F}(X)$ onto G, and that this map is a homomorphism, follows directly from the definitions. This completes the proof.

The elements of the kernel R of the epimorphism $\mathbf{F}(X) \to G$, are called the *relators* (or sometimes *relations*) of the group G, in terms of the alphabet X. If a subset R_1 of these relators is such that the smallest normal subgroup containing R_1 (called the *normal closure* of R_1 in $\mathbf{F}(X)$) is R itself, then we call R_1 a *set of defining relators* in the alphabet X. Since $G \simeq \mathbf{F}(X)/R$, the alphabet X and the set R_1 of words completely determines G (up to isomorphism of course). We shall call the pair $\langle X \mid R_1 \rangle$ (written that way), a *presentation of the group G in terms of generators and relations*, or, more briefly, a *presentation* of G, and we shall write $G \simeq \langle X \mid R_1 \rangle$. This construction is due to von Dyck (1856–1934). Of course a single group can be given many different presentations; the merit of one or the other presentation will depend on the particular problem at issue. Of special interest are those groups which have a finite presentation, i.e. for which finite X and R_1 can be found; such groups are called *finitely presented*.

14.1.6. Exercise. $\mathbf{C}_2 \times \mathbf{C}_2 \simeq \langle x, y \mid x^2, y^2, x^{-1}y^{-1}xy \rangle$.

14.1.7. Exercise. The free abelian group of rank n has the presentation

$$\langle x_1, \ldots, x_n \mid x_i^{-1}x_j^{-1}x_ix_j, \ 1 \leq i < j \leq n \rangle.$$

14.1.8. Exercise. $\mathbf{S}_3 \simeq \langle x, y \mid x^2, y^3, (xy)^2 \rangle$.

14.1.9. Exercise. Find presentations for the groups of order pq, where p, q are primes and $p < q$.

14.1.10. Exercise. The group of all matrices of the form $\left(\begin{smallmatrix} \varepsilon & * \\ 0 & 1 \end{smallmatrix}\right)$, $\varepsilon = \pm 1$, over the ring $\mathbf{Z}_n$ has the presentation $\langle x, y \mid x^2, y^n, (xy)^2 \rangle$ (the so-called "dihedral group of order $2n$": it is isomorphic to the group of symmetries of a regular n-gon).

14.1.11. Exercise. The subgroup of $\mathbf{F}(x, y)$ generated by the elements $x^n yx^n$, $n = 0, 1, 2, \ldots$, is freely generated by them (i.e. can be presented in terms of these generators with the empty set of relators).

The groups that arise in topology (and in some other areas) occur naturally in terms of presentations (for instance in knot theory). Details of the theory of group presentations may be found in [9, 27].

14.2. A Matrix Representation

In this section we shall show that any countable free group can be embedded in $\mathbf{SL}_2(\mathbf{Z})$. Since the groups $\mathbf{F}_1, \mathbf{F}_2, \ldots$ can be embedded in $\mathbf{F}_\infty$, and $\mathbf{F}_\infty$ in $\mathbf{F}_2$ (by Exercise 14.1.11 above), it suffices to display in $\mathbf{SL}_2(\mathbf{Z})$ a subgroup free of rank 2.

14.2.1. Theorem. *Let n be an integer ≥ 2. The subgroup of $\mathbf{SL}_2(\mathbf{Z})$ generated by the transvections*

$$t_{12}(n) = \begin{pmatrix} 1 & n \\ 0 & 1 \end{pmatrix}, \qquad t_{21}(n) = \begin{pmatrix} 1 & 0 \\ n & 1 \end{pmatrix},$$

is freely generated by them; i.e. there are no nontrivial relations on these two elements.

PROOF. For the sake of brevity we write $a = t_{12}(n)$, $b = t_{21}(n)$. Let w be an alternating product of nonzero powers of a and b. We wish to show that $w \neq e$. If w begins with a power of b then conjugate w by this power of b, and consider instead the resulting product, which now, of course, begins with a power of a (unless w itself was just a nonzero power of b: but in this case w is easily seen to be nontrivial). Thus we may suppose that

$$w = a^{\alpha_1} b^{\alpha_2} \cdots c^{\alpha_r}, \quad \text{where } c = a \text{ or } b, \text{ and all } \alpha_i \neq 0.$$

Let z_i denote the first row of the matrix $a^{\alpha_1} b^{\alpha_2} \cdots c^{\alpha_i}$. If $z_{2k-1} = (x_{2k-1}, x_{2k})$, then

$$z_{2k} = z_{2k-1} b^{\alpha_{2k}} = (x_{2k+1}, x_{2k}),$$

$$z_{2k+1} = z_{2k} a^{\alpha_{2k+1}} = (x_{2k+1}, x_{2k+2}),$$

where $x_{2k+1} = x_{2k-1} + n\alpha_{2k} x_{2k}$, and $x_{2k+2} = x_{2k} + n\alpha_{2k+1} x_{2k+1}$. The last two equations can be combined in the single equation

$$x_{i+2} = x_i + n\alpha_{i+1} x_{i+1}, \quad \text{for } i = 1, 2, \ldots, r-1.$$

The theorem will follow if we can show that the integer $|x_i|$ increases as i goes from 1 to $r+1$. For $i = 1, 2$ this can be verified directly. The inductive step is then as follows:

$$|x_{i+2}| \geq n|\alpha_{i+1}| \, |x_{i+1}| - |x_i| \geq 2|x_{i+1}| - |x_i| \geq |x_{i+1}| + 1.$$

This completes the proof.

The embeddings $\mathbf{F}_2 \to \mathbf{SL}_2(\mathbf{Z})$, afforded by this theorem, are extremely useful as a tool for investigating the properties of free groups. We illustrate this with the following application.

14.2.2. Theorem. *Let p be any prime. Every free group $\mathbf{F}(X)$ is residually a finite p-group.*

PROOF. Let $\hat{X}$ be a finite subset of X. Mapping the free generators outside $\hat{X}$ to the identity element, we obtain an epimorphism $\mathbf{F}(X) \to \mathbf{F}(\hat{X})$. Since the kernels of all such homomorphisms (obtained by letting $\hat{X}$ range over all finite subsets of X) have trivial intersection, we have by Remak's theorem (4.3.9) that $\mathbf{F}(X)$ is a subdirect product of countable free groups. Hence we may assume that $\mathbf{F}(X)$ is countable, and so, by Theorem 14.2.1, embeddable in the congruence subgroup $\Gamma_2(n)$. Recall that by definition (see Exercise 4.2.7):

$$\Gamma_2(n) = \{x \mid x \in \mathbf{SL}_2(\mathbf{Z}), \, x \equiv e \, (\text{mod } n)\}.$$

Being the kernels of the obvious homomorphisms $\mathbf{SL}_2(\mathbf{Z}) \to \mathbf{SL}_2(\mathbf{Z}_{p^k})$, the subgroups $\Gamma_2(p^k)$ are normal and of finite index in $\mathbf{SL}_2(\mathbf{Z})$; furthermore the intersection of the subgroups $\Gamma_2(p^k)$, $k = 1, 2, \ldots$, is clearly trivial. Hence, again by Remak's theorem, $\Gamma_2(p)$ is a subdirect product of the finite groups $\Gamma_2(p)/\Gamma_2(p^k)$, $k = 1, 2, \ldots$. Finally, for any $z \in \mathbf{SL}_2(\mathbf{Z})$,

$$(e + pz)^p = \sum_{i=0}^{p} \binom{p}{i} (pz)^i \equiv e \, (\text{mod } p^2),$$

$$(e + p^2 z)^p = \sum_{i=0}^{p} \binom{p}{i} (p^2 z)^i \equiv e \, (\text{mod } p^3),$$

$$\cdots \cdots \cdots \cdots \cdots \cdots \cdots \cdots \cdots ,$$

so that $\Gamma_2(p)/\Gamma_2(p^k)$ is a p-group, and the theorem is proved.

It is especially noteworthy that in this proof the residual finiteness of free groups was deduced from the residual finiteness of $\mathbf{GL}_2(\mathbf{Z})$. This is a particular instance of a more general phenomenon: we can often infer the residual finiteness of a group from the knowledge that it can be represented (faithfully) by matrices, thanks to the following theorem of A. I. Mal'cev [Matem. sb. **8** (1940), 405–421]: *A finitely generated group of matrices over a field is residually finite.* A simple proof of this theorem may be found in [30] Part 3, p. 99.

It is easy to check that the set

$$G_n = \{x \mid x \in \mathbf{SL}_2(\mathbf{Z}), \, x_{11} \equiv x_{22} \equiv 1 (\text{mod } n^2), \, x_{12} \equiv x_{21} \equiv 0 (\text{mod } n)\}$$

is a subgroup of $\mathbf{SL}_2(\mathbf{Z})$, and that

$$\langle t_{12}(n), t_{21}(n) \rangle \leq G_n.$$

Though in general we do not have equality here, I. N. Sanov [Doklady Akad. Nauk SSSR **57** (1947), 657–659] has shown that for $n = 2$ we do:

$$\langle t_{12}(2), t_{21}(2) \rangle = G_2.$$

In other words every matrix from G_2 can be expressed as a product of powers of the transvections $a = t_{12}(2)$ and $b = t_{21}(2)$ (and by Theorem 14.2.1 this expression will be unique). The way in which such a matrix can be decomposed explicitly as such a product is suggested by the following equations:

$$\begin{pmatrix} x_{11} & x_{12} \\ x_{21} & x_{22} \end{pmatrix} a^\alpha = \begin{pmatrix} x_{11} & x_{12}+2\alpha x_{11} \\ x_{21} & x_{22}+2\alpha x_{21} \end{pmatrix},$$

$$\begin{pmatrix} x_{11} & x_{12} \\ x_{21} & x_{22} \end{pmatrix} b^\beta = \begin{pmatrix} x_{11}+2\beta x_{12} & x_{12} \\ x_{21}+2\beta x_{22} & x_{22} \end{pmatrix}.$$

We shall restrict ourselves to illustrating the general procedure by carrying it out for the particular matrix

$$x = \begin{pmatrix} -23 & -86 \\ -4 & -15 \end{pmatrix}.$$

Here $|x_{12}| > |x_{11}|$. We look for an integer α such that in the matrix $x' = xa^\alpha$ we have $|x'_{12}| < |x'_{11}|$. Thus we divide $x_{12}+|x_{11}|$ by $2x_{11}$: explicitly, $-63 = -46\cdot 2+29$, and take $\alpha = -2$ (the quotient with its sign changed). Then

$$x' = xa^{-2} = \begin{pmatrix} -23 & 6 \\ -4 & 1 \end{pmatrix}.$$

Next we look for an integer β such that in the matrix $x'' = x'b^\beta$ we have $|x''_{12}| > |x''_{11}|$. We divide $x'_{11}+|x'_{12}|$ by $2x'_{12}$ and obtain: $-17 = 12\cdot(-2)+7$. We choose $\beta = 2$ (the quotient with changed sign), and we then have

$$x'' = x'b^2 = \begin{pmatrix} 1 & 6 \\ 0 & 1 \end{pmatrix} = a^3.$$

Hence $x = a^3 b^{-2} a^2$.

14.2.3. Exercise. Describe and justify the algorithm for decomposing matrices from G_2 as products of the generators $t_{12}(2)$, $t_{21}(2)$. Using this algorithm, decompose the matrix $\begin{pmatrix} 321 & -86 \\ 698 & -187 \end{pmatrix}$.

14.3. Subgroups

The subgroups of a free group are, as it happens, also free. This was proved by J. Nielsen in 1921 for finitely generated free groups, and in full generality by O. Schreier in 1927. In this section we shall describe Schreier's method, which has the advantage that it often enables us to write down immediately a set of free generators for a subgroup of a free group.

Let H be a subgroup of an arbitrary group G, and choose an element from each right coset of H in G, making the choice e for the coset H. (In other words choose a right transversal for H in G containing e.) The function $\phi: G \to G$, which has a constant value on each right coset, namely the chosen representative of that coset, is called the *choice* or *right coset representative function* corresponding to the transversal. The following properties of this function are immediate: $(u\phi)\phi = u\phi$ and $[(u\phi)v]\phi = (uv)\phi$, where $u, v \in G$.

14.3.1. Theorem. *Let M be a generating set for an arbitrary group G, let H be a subgroup of G, and let ϕ be a right coset representative function corresponding to a right transversal T for H in G. Then*

$$H = \langle tx[(tx)\phi]^{-1} \mid t \in T, x \in M \rangle.$$

PROOF. It is clear that the elements $tx[(tx)\phi]^{-1}$ lie in H. We have to show that any element of H can be expressed as a product of these elements and their inverses. A direct verification gives that

$$(tx[(tx)\phi]^{-1})^{-1} = \hat{t}x^{-1}[(\hat{t}x^{-1})\phi]^{-1}, \quad \text{where } \hat{t} = (tx)\phi.$$

Note that $(\hat{t}x^{-1})\phi = t$. Now let $u = x_1^{\varepsilon_1} \cdots x_r^{\varepsilon_r}$, $x_i \in M$, $\varepsilon_i = \pm 1$, be any element of H. We shall "rewrite" u in terms of the $tx[(tx)\phi]^{-1}$. If we define $u_1 = 1$, $u_{i+1} = x_1^{\varepsilon_1} \cdots x_i^{\varepsilon_i}$, $i = 1, \ldots, r-1$, then the rewriting proceeds by the following steps:

$$u = u_1 x_1^{\varepsilon_1}[(u_1 x_1^{\varepsilon_1})\phi]^{-1} \cdot u_2\phi \cdot x_2^{\varepsilon_2} \cdots x_r^{\varepsilon_r}, \quad \text{since } u_2\phi = (u_1 x_1^{\varepsilon_1})\phi;$$

$$= u_1 x_1^{\varepsilon_1}[(u_1 x_1^{\varepsilon_1})\phi]^{-1} \cdot u_2\phi x_2^{\varepsilon_2}[(u_2 x_2^{\varepsilon_2})\phi]^{-1} \cdot u_3\phi \cdot x_3^{\varepsilon_3} \cdots x_r^{\varepsilon_r},$$

$$\text{since } u_3\phi = (u_2 x_2^{\varepsilon_2})\phi;$$

. . . .

Continuing in this way we get finally

$$u = u_1 x_1^{\varepsilon_1}[(u_1 x_1^{\varepsilon_1})\phi]^{-1} \cdot u_2\phi x_2^{\varepsilon_2}[(u_2 x_2^{\varepsilon_2})\phi]^{-1} \cdots u_r\phi x_r^{\varepsilon_r}[(u_r x_r^{\varepsilon_r})\phi]^{-1}. \tag{1}$$

We have used throughout the fact that

$$u_{i+1}\phi = (u_i x_i^{\varepsilon_i})\phi, \qquad i = 1, \ldots, r-1,$$

and, in the last step, that $(u_r x_r^{\varepsilon_r})\phi = u\phi = e$. The equation (1) expresses u as a product of r elements of the required form, completing the proof.

14.3.2. Exercise. A subgroup of finite index in a finitely generated group, is finitely generated.

14.3.3. Exercise. In the free group $\mathbf{F}_n$ the words of even length form a subgroup. Find generators for this subgroup.

14.3.4. Exercise. Using the proof of Theorem 14.3.1 above, and the generators for $\mathbf{S}_n$ given in Exercise 2.2.4, write down generators for $\mathbf{A}_n$.

A (right) *Schreier system* in a free group is a set of words which, in reduced form, have the property that all their initial segments are also in the set; i.e. the set is closed under taking initial segments. If a Schreier system happens to be also a right transversal for a subgroup H of the free group, then it is called a *Schreier transversal* for H.

14.3.5. Theorem (Nielsen–Schreier). *Let H be an arbitrary subgroup of the free group $F = \mathbf{F}(X)$, where X is any alphabet. Then there exists at least one Schreier transversal T for H in F. Further, if ϕ is the corresponding coset representative function, then H is freely generated by those of the elements $tx[(tx)\phi]^{-1}$ that are nontrivial, where t ranges over T, and x over X.*

PROOF. (i) We first prove the existence of a Schreier transversal. By the *length of a right coset* of H in F, we shall mean the length of a shortest word in the coset. We shall construct a Schreier transversal using induction on the length of a coset.

Choose the empty word as the representative of the coset H. From each coset of length 1 choose as representative an element of length 1. More generally, suppose that for cosets of length $<r$ (where $r \geq 1$), representatives have already been chosen in such a way that the length of each representative is the same as the length of its coset. Thus ϕ is assumed to be defined on all cosets of length $<r$. Let Hg be any coset of length r, where we may suppose that g has length r, say $g = x_1 \cdots x_r$, $x_i \in X \cup X^{-1}$. Then as the representative $g\phi$ of Hg choose $(x_1 \ldots x_{r-1})\phi \cdot x_r$. The inductive step is completed by doing this for all cosets of length r. It is clear that the resulting transversal is a Schreier system.

(ii) Let T be a Schreier transversal for H in F and let ϕ be the corresponding coset representative function. We wish to show that the nontrivial elements among those of the form

$$tx[(tx)\phi]^{-1}, \qquad t \in T, \quad x \in X, \tag{2}$$

freely generate H. We already know (from Theorem 14.3.1 above) that these elements generate H. It remains to prove that there is no nontrivial relation among them.

As the first step we show that each nontrivial word in (2) is reduced as written, assuming that the t are in reduced form. To see this note first that any cancellation in (2) must begin with the symbol x, so that for cancellation to occur either $t \equiv t_1 x^{-1}$, in which case $tx[(tx)\phi]^{-1} = t_1(t_1\phi)^{-1} = t_1 t_1^{-1} = e$; or $[(tx)\phi]^{-1} \equiv x^{-1}t_2^{-1}$, whence $t = t_2$ and again $tx[(tx)\phi]^{-1} = e$.

The second step involves the following. Let u, v be nontrivial words of the form (2) or inverses of such words, such that $uv \neq e$. From the proof of Theorem 14.3.1 above, and the above argument, we know that, in reduced form,

$$u = tx^{\varepsilon}[(tx^{\varepsilon})\phi]^{-1}, \qquad v = sy^{\delta}[(sy^{\delta})\phi]^{-1},$$

$$t, s \in T, \qquad x, y \in X, \qquad \varepsilon, \delta = \pm 1.$$

Since u and v are reduced as here written, any cancellation in the product uv must begin at the "interface" between them. The crucial point is that any such cancellation halts before reaching the symbol x^{ε} in u and y^{δ} in v. For if x^{ε} were to cancel earlier than y^{δ}, we should have that $s \equiv t_1 x^{-\varepsilon}w$, where

$t_1 \equiv (tx^\varepsilon)\phi$, whence, since T is a Schreier system,

$$t_1 x^{-\varepsilon} \equiv (t_1 x^{-\varepsilon})\phi = t.$$

But this means that $u = 1$, a contradiction. If y^δ were to cancel earlier than x^ε then we should have that $t_1 \equiv sy^\delta v$, where $t_1 \equiv (tx^\varepsilon)\phi$, whence

$$sy^\delta = (sy^\delta)\phi.$$

This is impossible since $v \neq 1$. Finally the simultaneous cancellation of x^ε and y^δ is impossible since $uv \neq 1$.

For the third and final step, suppose given a word in the nontrivial elements of the form (2), which is reduced as a word in these elements. We wish to show that as a word in the alphabet X it does not reduce to the empty word. This is now quite easy since by the first two steps in the argument any cancellation must begin at the interfaces between adjacent elements (2), and must halt before reaching the "core" symbols x.

14.3.6. Exercise. In $\mathbf{F}(x, y)$ find a Schreier transversal and thence free generators, for the commutator subgroup. Do the same thing for the normal closure of the elements x, y^r, r an integer.

The next theorem shows how to compute the rank of a subgroup of finite index in a free group of finite rank.

14.3.7. Theorem. *Let* $H \leq \mathbf{F}_n = \mathbf{F}(x_1, \ldots, x_n)$, *have finite index j in* $\mathbf{F}_n$. *Then H has rank*

$$m = 1 + (n - 1)j.$$

PROOF. We use the notation of the preceding theorem (with $X = \{x_1, \ldots, x_n\}$). Let M denote the set of all formal expressions (2): there are nj such expressions. We have to discover which of these define the identity element. With this aim in mind write T_0 for $T \setminus \{e\}$, and define a map $\tau \colon T_0 \to M$, by

$$t\tau = \begin{cases} t'x[(t'x)\phi]^{-1} & \text{for } t \equiv t'x, \; x \in X, \\ tx[(tx)\phi]^{-1} & \text{for } t \equiv t'x^{-1}, \; x \in X. \end{cases}$$

It is straightforward to check that τ is one-to-one, and that $T_0\tau$ consists of just those expressions in M which define the identity element. Since the rank of H is the number of expressions remaining, namely $nj - (j - 1)$, the theorem is proved.

14.3.8. Exercise. A subgroup of finite index in a free group of infinite rank, also has infinite rank (i.e. the condition that n be finite may be removed from Theorem 14.3.7 without affecting the conclusion).

14.3.9. Exercise. If F_1, F_2 are free normal subgroups of the same finite index in a group G, then $F_1 \simeq F_2$.

14.4. The Lower Central Series and the Derived Series

Let G be an arbitrary group. We define a certain descending chain

$$\gamma_1 G \geq \gamma_2 G \geq \cdots \tag{3}$$

of subgroups of G, in the following manner: Put $\gamma_1 G = G$, and define $\gamma_{n+1} G$ in terms of $\gamma_n G$ by setting $\gamma_{n+1} G = [\gamma_n G, G]$. (Recall that $[A, B]$ is the mutual commutator of A and B (Ch. 1, §3.2).) The resulting chain (3) is called the *lower central series* of G, and $\gamma_n G$ its *nth term*.

The *left-normed commutator* $[x_1, \ldots, x_n]$ *of weight* n in $x_1, \ldots, x_n$, is defined for $n = 1$ and 2 to be respectively x_1 and $[x_1, x_2]$, while for $n > 2$ it is defined inductively by

$$[x_1, \ldots, x_n] = [[x_1, \ldots, x_{n-1}], x_n].$$

14.4.1. Exercise. For every group G

$$\gamma_n G = \langle [g_1, \ldots, g_n] \mid g_i \in G \rangle, \qquad n = 1, 2, \ldots.$$

14.4.2. Exercise. For every group G

$$[\gamma_i G, \gamma_j G] \leq \gamma_{i+j} G, \qquad i, j = 1, 2, \ldots.$$

(Hint. Use induction and the "lemma on the three commutators" (3.2.10).)

14.4.3. Exercise. For any group G the nth derived subgroup is contained in the 2^n-th member of the lower central series; the nth member of the lower central series of a subgroup is contained in the nth member of the lower central series of G.

14.4.4. Theorem (Magnus). *The terms of the lower central series of any free group F have trivial intersection; i.e.*

$$\bigcap_{i=1}^{\infty} \gamma_i F = 1.$$

PROOF. (i) Suppose first that F is countable. Then by Theorem 14.2.1 above F is embeddable in the congruence subgroup $\Gamma_2(n)$, $n \geq 2$, so that it suffices to show that the intersection of the lower central series of $\Gamma_2(n)$ is trivial. Consider matrices

$$g = e + n^k a \in \Gamma_2(n^k), \quad \text{and} \quad f = e + n^l b \in \Gamma_2(n^l).$$

Write $g^{-1} = e + n^k a'$, and $f^{-1} = e + n^l b'$. Then the commutator

$$[g, f] = (e + n^k a')(e + n^l b')(e + n^k a)(e + n^l b)$$

$$= e + (n^k a' + n^k a + n^{2k} a' a) + (n^l b' + n^l b + n^{2l} b' b) + c,$$

where c is congruent to the zero matrix modulo n^{k+l}. Now from $g^{-1} g = e = f^{-1} f$, we get that $(n^k a' + n^k a + n^{2k} a' a)$ and $(n^l b' + n^l b + n^{2l} b' b)$ are both the zero matrix. Hence $[g, f] \in \Gamma_2(n^{k+l})$.

From this computation we deduce that

$$[\Gamma_2(n^k), \Gamma_2(n^l)] \leq \Gamma_2(n^{k+l}),$$

whence $\gamma_i \Gamma_2(n) \leq \Gamma_2(n^i)$. Since $\bigcap_i \Gamma_2(n^i) = 1$, the result follows for countable F.

(ii) Now let F be any free group, with a free basis X. Suppose that there is a nontrivial element f which is contained in all $\gamma_i F$. For each $i = 1, 2, \ldots$, express f as a product of left-normed commutators of weight i, and denote by S_i the set of entries in these commutators. The set $\bigcup_{i=1}^{\infty} S_i$, being itself countable, generates a countable subgroup F, which is of course free by the Nielsen–Schreier theorem. However, it has the element f in every term of its lower central series, contradicting Part (i) of the proof.

Note that by Exercise 14.4.3 and Magnus' theorem, the derived series of a free group also has trivial intersection. (In other words, the ωth term of the derived series is trivial.)

Let w be a word in the alphabet X. Define the *logarithm of w to the base $x \in X$*, or $\log_x w$, to be the sum of the exponents on the letter x wherever it occurs in w. (One meets frequently with the alternative terminology "exponent sum on x in w.") Clearly $\log_x w$ is an integer and has the same value for equivalent words. It is also fairly clear that the map sending each $w \in \mathbf{F}(X)$ to its family $(\log_x w)_{x \in X}$ of logarithms, is an epimorphism from $\mathbf{F}(X)$ onto a free *abelian* group of the same rank. It is not difficult to see that a word has its logarithms to all bases zero if and only if it is in the derived group, so that the kernel of the above epimorphism is just $[\mathbf{F}(X), \mathbf{F}(X)]$. By applying this reasoning to each term of the derived series of a free group we obtain the following result:

The factors of the derived series of a free group are free abelian.

We note without proof the fact that the factors of the lower central series of a free group are also free abelian (Witt's theorem). The means required for proving this are considerably more subtle.

14.4.5. Exercise. Free groups are isomorphic if and only if they have the same rank.

14.4.6. Exercise. Using the commutator identities of §3.2, show that the factors of the lower central series of a free group of finite rank are finitely generated abelian groups.

§15. Varieties

A subclass of groups distinguished from the general population of groups by means of identical relations, or laws, is called a *variety*. (The name was appropriated from the lexicon of algebraic geometry.) For example the class of all abelian groups is a variety, since its members are just those groups satisfying the law $xy = yx$. There is a close connexion between varieties and free groups, since laws are just elements of free groups (rewrite $xy = yx$ in the form $x^{-1}y^{-1}xy = 1$). A large number of papers has been written on the theory of varieties, and also a book ([31]).

Here we shall introduce the reader to the basic ideas and method of this theory; in particular we shall show that a variety may be defined equivalently as a class of groups closed under taking subgroups, epimorphic images and Cartesian products.

15.1. Laws and Varieties

A word v in the alphabet $x_1, x_2, \ldots$, is said to be a *law* in a class $\mathfrak{L}$ of groups if for every group G in $\mathfrak{L}$, v becomes trivial whatever values the arguments $x_1, x_2, \ldots$ are assigned from G; i.e. if $v = v(x_1, \ldots, x_n)$ then $v(g_1, \ldots, g_n) = 1$ for all $g_i \in G$, and for all $G \in \mathfrak{L}$. Let V be a set of words in the alphabet $x_1, x_2, \ldots$, and let G be a group. The values taken by the words in V as the arguments $x_1, x_2, \ldots$, run through G, are of course in general not all trivial. We call the subgroup $V(G)$ generated by these values the *verbal subgroup of G relative to V*. Thus

$$V(G) = \langle v(g_1, \ldots, g_{n(v)}) \,|\, v \in V, g_i \in G \rangle.$$

This subgroup measures the deviation, in some sense, of the group G from the groups of the variety $\mathfrak{B}$ defined by the laws V. As two obvious examples of verbal subgroups we mention the commutator subgroup and the dth power of a group: they are defined by the single words $[x, y]$ and x^d respectively, and may be thought of as measuring the deviation of the group from being abelian and having exponent d.

15.1.1. Exercise. The terms of the lower central series and the derived series are verbal subgroups.

15.1.2. Exercise. A verbal subgroup of a verbal subgroup of a group G is itself verbal in G.

15.1.3. Exercise. The verbal subgroups of $F = \mathbf{F}(X)$ are just the subsets H of that group with the following properties:

(i) $u, v \in H \Rightarrow uv^{-1} \in H$;
(ii) $u = u(x_1, \ldots, x_n) \in H$, $v_1, \ldots, v_n \in F \Rightarrow u(v_1, \ldots, v_n) \in H$.

We shall now establish the existence of, and incidentally describe, the groups free in a variety in the sense of the beginning of Chapter 3.

15.1.4. Theorem. *Let $\mathfrak{V}$ be a variety of groups defined by a set V of laws. For any alphabet $X = \{x_i \mid i \in I\}$, define*

$$\mathbf{F}_V(X) = \mathbf{F}(X) / V(\mathbf{F}(X)),$$

and denote by $\hat{x}$ the image of $x \in X$ under the natural epimorphism $\mathbf{F}(X) \rightarrow \mathbf{F}_V(X)$. If G is a group from $\mathfrak{V}$, and $g_i, i \in I$, are arbitrary elements of G, then the map $\hat{x}_i \rightarrow g_i$ extends (uniquely) to a homomorphism $\mathbf{F}_V(X) \rightarrow G$. In other words the groups $\mathbf{F}_V(X)$ are free in the variety $\mathfrak{V}$. Moreover every group free in $\mathfrak{V}$ is isomorphic to some $\mathbf{F}_V(X)$; i.e. they exhaust the groups free in $\mathfrak{V}$.

PROOF. (i) The map $x_i \rightarrow g_i$ extends to a unique homomorphism $\psi \colon \mathbf{F}(X) \rightarrow G$, by virtue of the "absolute" freeness of $\mathbf{F}(X)$. Since $G \in \mathfrak{V}$, the kernel of ψ contains $V(\mathbf{F}(X))$. A little computation in the spirit of the homomorphism theorems (§4.2) completes the proof that $\hat{x}_i \rightarrow g_i$ extends to a homomorphism.

(ii) Let F be free in $\mathfrak{V}$ with free generators $f_i, i \in I$. Then by the freeness in $\mathfrak{V}$ of F, the map $f_i \rightarrow \hat{x}_i$ extends to an epimorphism $F \rightarrow \mathbf{F}_V(X)$. But by the freeness in $\mathfrak{V}$ of $\mathbf{F}_V(X)$, the map $\hat{x}_i \rightarrow f_i$ also extends to an epimorphism. Thus $F \simeq \mathbf{F}_V(X)$, and the proof is complete.

The set $\{\hat{x}_i \mid i \in I\}$ is called a *free basis* or *free generating set* for the group $\mathbf{F}_V(X)$, and the cardinal of the free basis is the *rank* of $\mathbf{F}_V(X)$. The synonymous concepts for free and free abelian groups, which we met with earlier, now fall into place as particular instances of these general concepts. The free group of $\mathfrak{V}$ of rank $\mathfrak{n}$, for any cardinal $\mathfrak{n}$, is also denoted by $\mathbf{F}_\mathfrak{n}(\mathfrak{V})$.

In §7 we saw that the free abelian groups are simple enough for us to be able to described them exhaustively. The free groups of other varieties may be much more complicated. For instance it was completely unknown until quite recently whether or not the finite rank free groups in the variety defined by the single law $x^d = 1$, $d \geq 5$, are finite. This famous problem, known as "Burnside's problem," was finally settled by P. S. Novikov and S. I. Adjan [Izv. Akad. Nauk SSSR, ser. matem., **32**, Nos. 1, 2, 3 (1968)]. They showed that for all odd $d \geq 4381$, $\mathbf{B}(n, d)$, the free group of rank $n \geq 2$ in the variety of all groups of exponent d, is infinite. Later the restriction "for all odd $d \geq 4381$" was weakened to "for all odd $d \geq 665$" ([1]). It turns out that all the abelian subgroups of $\mathbf{B}(n, d)$ are cyclic, and that its center is trivial ([1]). It is appropriate to mention here also the deep result of A. I. Kostrikin [Izv. Akad. Nauk SSSR, ser. matem., **23** (1959), 3–34] that for each natural number n and each prime p there are only a finite number of finite n-generator groups of exponent p.

15.1.5. Exercise. The quaternion group $\langle x, y \mid x^4, x^2 y^2, x^{-1} yxy \rangle$ is not free in any variety.

For any word $w = w(x_1, \ldots, x_n)$ and any homomorphism $\phi : G \to G^*$, it is obvious that

$$w(g_1, \ldots, g_n)^\phi = w(g_1^\phi, \ldots, g_n^\phi), \qquad g_i \in G.$$

hence the homomorphic image of a verbal subgroup $V(G)$ is contained in $V(G^*)$. In particular the verbal subgroups of a group are fully invariant.

15.1.6. Exercise. Find a subgroup of $C_2 \times C_4$ which is fully invariant, but not verbal.

For groups free in some variety the situation is more pleasant:

15.1.7. Theorem. *If a group G is free in some variety $\mathfrak{V}$, then every fully invariant subgroup of G is verbal.*

PROOF. Suppose that H is a fully invariant subgroup of G. Write W for the set of words $w(x_1, \ldots, x_n)$ of F_∞ with the property that $w(g_1, \ldots, g_n) \in H$ for all $g_1, \ldots, g_n \in G$. We shall show that $H = W(G)$. It is clear from the definition of W that $H \geq W(G)$. The reverse inclusion is not too difficult: write $h \in H$ as a word $w(f_{i_1}, \ldots, f_{i_n})$ in a set $\{f_i \mid i \in I\}$ of free generators of G; then the full invariance of H implies that $w(x_1, \ldots, x_n) \in W$, since any replacement of $x_1, \ldots, x_n$ by $g_1, \ldots, g_n \in G$, can be achieved by applying to $w(f_{i_1}, \ldots, f_{i_n})$ an endomorphism of G sending f_{i_1} to $g_1, \ldots, f_{i_n}$ to g_n. This completes the proof.

The correspondence between varieties and sets of words from F_∞ defining them is not one-to-one; it is easy to think of different sets of laws which define the same variety. The set of *all* laws satisfied simultaneously by all the groups in a given variety forms a verbal subgroup of F_∞: this is the appropriate subset to associate with the variety. It is easy to see that this correspondence of verbal subgroup of F_∞ with variety, is one-to-one, so that the study of varieties is equivalent to the study of the verbal subgroups of F_∞.

15.1.8. Exercise. The subgroup generated by a set of verbal subgroups of F_∞ is again verbal. What variety does it define?

Two sets V, W of laws are said to be *equivalent* if they define the same variety, or, in other words, if $V(F_\infty) = W(F_\infty)$.

15.1.9. Exercise. Any finite set of laws is equivalent to a single law.

The question of the existence of varieties not defined by finite sets of laws, remained unanswered for more than 20 years. It was finally settled in the affirmative by A. Ju. Ol'šanskiĭ, who in fact showed that there are continuously many varieties of groups. Specific examples of varieties not defined by finite sets of laws were first given by S. I. Adjan [Izv. Akad. Nauk

SSSR, ser. matem., **34**, No. 4 (1970), 715–734], and M. R. Vaughan-Lee [Bull. London Math. Soc. **2** (1970), 280–286].

15.1.10. Theorem (B. H. Neumann). *Every set V of words in the symbols $x_1, x_2, \ldots$, is equivalent to a set of the form $W = \{x_1^d, u_1, u_2, \ldots\}$, where $d \geq 0$, and the u_i are in the commutator subgroup of $\mathbf{F}(x_1, x_2, \ldots) = \mathbf{F}_\infty$.*

PROOF. Write each $v \in V$ in the form

$$v = x_{i_1}^{n_1} \cdots x_{i_s}^{n_s} u,$$

where $s \geq 0$, the subscripts $i_1, \ldots, i_s$ are all different, the n_i are all non-zero, and u belongs to the commutator subgroup of $\mathbf{F}_\infty$. Let d be the highest common factor of $n_1, \ldots, n_s$ taken over all v in V (and take $d = 0$ if $s = 0$ for all v in V, i.e. if all v lie in $\mathbf{F}'_\infty$). The non-negative integer d and the elements u are what we were looking for. Let us now prove this. It is not too difficult to see that $V(\mathbf{F}_\infty) \leq W(\mathbf{F}_\infty)$. For the reverse inclusion note first that $x_{i_k}^{n_k}$ is in $V(\mathbf{F}_\infty)$, since it can be obtained from v by replacing all the other x_i occurring in it by the identity element. Hence $x_1^{n_k} \in V(\mathbf{F}_\infty)$, so that finally $x_1^d \in V(\mathbf{F}_\infty)$, and then, of course, the elements u will also lie in $V(\mathbf{F}_\infty)$.

15.2. An Alternative Approach to Varieties

For any class $\mathfrak{L}$ of groups, we denote by $\mathrm{s}\mathfrak{L}$, $\mathrm{Q}\mathfrak{L}$, $\mathrm{c}\mathfrak{L}$ the respective closures of $\mathfrak{L}$ under taking subgroups, homomorphic images (Q for "quotient"), and Cartesian products, of groups in $\mathfrak{L}$. If $\mathfrak{L}$ happens to be a variety then clearly $\mathrm{s}\mathfrak{L} = \mathfrak{L}$, $\mathrm{Q}\mathfrak{L} = \mathfrak{L}$ and $\mathrm{c}\mathfrak{L} = \mathfrak{L}$. The converse of this is also true.

15.2.1. Theorem (Birkhoff). *Varieties are precisely those classes of groups that are closed under taking subgroups, homomorphic images and Cartesian products.*

PROOF. We wish to show that if $\mathrm{s}\mathfrak{L} = \mathfrak{L}$, $\mathrm{Q}\mathfrak{L} = \mathfrak{L}$ and $\mathrm{c}\mathfrak{L} = \mathfrak{L}$, then the class $\mathfrak{L}$ is a variety. Let V be the set of all laws satisfied by every group in $\mathfrak{L}$, and let $\mathfrak{B}$ be the variety defined by these laws. It is clear that $\mathfrak{L} \subseteq \mathfrak{B}$; we wish to establish that $\mathfrak{B} \subseteq \mathfrak{L}$. Since $\mathrm{Q}\mathfrak{L} = \mathfrak{L}$ it suffices to show that every group F free in $\mathfrak{B}$, belongs to $\mathfrak{L}$. We shall actually construct a Cartesian product of groups from $\mathfrak{L}$ which embeds F; this will imply that $F \in \mathfrak{L}$ since $\mathrm{c}\mathfrak{L} = \mathfrak{L}$.

By Theorem 15.1.4 above

$$F = \mathbf{F}(X)/V(\mathbf{F}(X))$$

for some alphabet $X = \{x_i \mid i \in I\}$. For each word $w(x_1, \ldots, x_n)$ not belonging to the set V, and each ordered n-tuple $\mathbf{i} = (i_1, \ldots, i_n)$ of distinct elements of I, choose a group $G_{w,\mathbf{i}}$, and from it elements $g_{w,\mathbf{i}}^j$, $j \in I$, such that $G_{w,\mathbf{i}} \in \mathfrak{L}$ and

$$w(g_{w,\mathbf{i}}^{i_1}, \ldots, g_{w,\mathbf{i}}^{i_n}) \neq e.$$

Form the Cartesian product

$$\prod_{(w,\mathbf{i})} G_{w,\mathbf{i}},$$

over all $(w, \mathbf{i})$, $w \notin V$, and consider the subgroup of it generated by the functions f_j, $j \in I$, defined by $f_j(w, \mathbf{i}) = g_{w,\mathbf{i}}^j$. It is easy to see that the only relations among the f_j are the laws in V, so that the f_j, $j \in I$, generate a group isomorphic to F. This completes the proof.

This proof actually shows that for any class $\mathfrak{L}$ of groups the smallest variety containing $\mathfrak{L}$ is $\mathrm{QSC}\mathfrak{L}$.

15.2.2. Exercise. Let G be a finite group and $\mathfrak{B}$ the variety defined by all the laws of G. Show that the finitely generated groups in $\mathfrak{B}$ are finite. (Hint. Use the comment preceding this exercise, with $\mathfrak{L} = \{G\}$.)

15.2.3. Exercise. In a finitely generated group G every subgroup of finite index contains a verbal subgroup of finite index.

(Solution. By Exercise 2.5.13 we may suppose that H is normal in G. Let V be the set of laws of the finite group G/H, and $\mathfrak{B}$ the variety defined by these laws. Since $G/V(G)$ is finitely generated and belongs to $\mathfrak{B}$, it also is finite (by the preceding Exercise 15.2.2).)

15.2.4. Exercise. Find classes of groups $\mathfrak{L}$, $\mathfrak{M}$, $\mathfrak{N}$ satisfying the following conditions:

$$\begin{array}{ccc}
\mathrm{s}\mathfrak{L} \neq \mathfrak{L}, & \mathrm{Q}\mathfrak{L} = \mathfrak{L}, & \mathrm{c}\mathfrak{L} = \mathfrak{L}, \\
\mathrm{s}\mathfrak{M} = \mathfrak{M}, & \mathrm{Q}\mathfrak{M} \neq \mathfrak{M}, & \mathrm{c}\mathfrak{M} = \mathfrak{M}, \\
\mathrm{s}\mathfrak{N} = \mathfrak{N}, & \mathrm{Q}\mathfrak{N} = \mathfrak{N}, & \mathrm{c}\mathfrak{N} \neq \mathfrak{N}.
\end{array}$$

Nilpotent Groups 6

In two earlier chapters we gained some familiarity with abelian and finite groups. Although, of course, these by no means account for all groups, every group is connected with them in some way or other. For instance the free groups of rank ≥ 2 are not abelian and not finite, yet, as we have seen, they possess descending normal chains with abelian factors, and they are residually finite. There are a great many papers on group theory devoted to establishing connexions such as these with finiteness and commutativity. As a result there exists a welter of conditions generalizing the conditions of finiteness or abelianness, ranging from the sublime to the ridiculous.

The most important generalizations of commutativity are solubility and nilpotency. Soluble groups are those that can be constructed from abelian groups by means of a finite number of successive extensions. They are especially well known for their relevance to the problem of solving algebraic equations by radicals (see the Introduction), whence their name.

Nilpotent groups form a class smaller than that of soluble groups but larger than that of abelian groups. Their definition is more complicated, but they can be more intimately studied than soluble groups.

This chapter is devoted to nilpotent groups and the next to soluble groups. As to generalizations of finiteness, we have already met with some of them—for instance periodicity, the property of being finitely generated—and others will be introduced as the need arises.

Many further interesting facts about nilpotent and soluble groups may be found in [8, 18, 20, 35, 42].

§16. General Properties and Examples

16.1. Definition

Let G be a group. A normal series

$$1 = G_0 \leq G_1 \leq \cdots \leq G_s = G \tag{1}$$

is said to be *central* if all of its factors are central; i.e. if

$$G_{i+1}/G_i \leq C(G/G_i), \quad \text{for all } i, \tag{2}$$

or equivalently, if

$$[G_{i+1}, G] \leq G_i, \quad \text{for all } i. \tag{3}$$

A group having a central series is called *nilpotent* (we shall explain the name below), and the length of the group's shortest central series is termed its *nilpotency class*. (The use of the word "class" for this number is unfortunate but firmly established; some adroitness is needed in order to avoid stylistic masterpieces like "class of nilpotent groups of a given class.") Our remark that the class of nilpotent groups is intermediate between the classes of abelian and soluble groups is now clear from the definition. Abelian groups are just the nilpotent groups of class ≤ 1.

16.1.1. Exercise. Any series satisfying condition (3) is automatically a normal series.

Let G be an arbitrary group. Guided by conditions (2) and (3) we attempt to construct central series for G: Define

$$\zeta_0 G = 1, \quad \zeta_{i+1}G/\zeta_i G = C(G/\zeta_i G), \quad i = 0, 1, 2, \ldots;$$

$$\gamma_1 G = G, \quad \gamma_{j+1}G = [\gamma_j G, G], \quad j = 1, 2, \ldots.$$

The subgroups $\zeta_i G$ are called the *higher centers* of the group G, while the groups $\gamma_i G$ are the familiar (from §14.4) terms of the *lower central series*:

$$G = \gamma_1 G \geq \gamma_2 G \geq \cdots;$$

the ascending series

$$1 = \zeta_0 G \leq \zeta_1 G \leq \cdots,$$

is the *upper central series* of G. (We remark in passing that it is easy to define $\zeta_\alpha G$ and $\gamma_\alpha G$ for any ordinal α; we shall however avoid this particular quagmire, at least for the time being.)

It is clear that if some higher center coincides with the whole group, or some term of the lower central series is trivial, then the group is nilpotent. Conversely, suppose that a group G is nilpotent and that (1) is an arbitrary central series for the group. For the sake of brevity write $Z_i = \zeta_i G$, $\Gamma_j = \gamma_j G$.

The definitions and an easy induction lead to the following scheme of inclusions:

$$1 = Z_0 \le Z_1 \le \cdots$$
$$\text{\scriptsize VI} \qquad \text{\scriptsize VI}$$
$$1 = G_0 \le G_1 \le \cdots \le G_{s-1} \le G_s = G \qquad (4)$$
$$\text{\scriptsize VI} \qquad \text{\scriptsize VI}$$
$$\cdots \le \ \Gamma_2 \ \le \Gamma_1 = G.$$

It is clear from these inclusions that both the upper and lower central series of a nilpotent group are finite, and that their lengths are equal to the nilpotency class of the group. (The scheme (4) makes especially clear the motivation for choosing the adjectives "upper" and "lower" for the two series.) Although the upper and lower central series of a nilpotent group have the same length, they themselves need not be the same (see Exercises 16.2.10 and 16.2.11 below).

16.1.2. EXAMPLE. Let K be a field. Using the formulae (9), (12) of 3.2.1, it is not difficult to see that, for $n \ge 3$, the series

$$\mathbf{UT}_n(K) = \mathbf{UT}_n^1(K) > \mathbf{UT}_n^2(K) > \cdots > \mathbf{UT}_n^n(K) = 1$$

is both the upper and lower central series of the group $\mathbf{UT}_n(K)$. Incidentally we get that this group is nilpotent of class $n-1$. One of the authors [Ju. I. Merzljakov, Central series and derived series of matrix groups, Algebra i Logika **3**, No. 4 (1964), 49–58] has computed central series for other matrix groups, in particular the lower central series of the principal congruence subgroups over local rings (see Exercise 4.2.7), and of the Sylow p-subgroups of $\mathbf{GL}_n(\mathbf{Z}_{p^m})$. In the latter case, for example, the situation can be described as follows. For brevity, we put $K = \mathbf{Z}_{p^m}$. Take the "empty" $n \times n$ matrix, i.e. a square divided into n^2 congruent smaller squares by lines parallel to its sides, and cover it with a "carpet" made up of the ideals K, pK, $p^2K, \ldots$, as shown in the diagram (where $n = 3$):

$$
\begin{array}{ccccc|ccc|ccc}
\cdots & p^2K & p^2K & p^2K & pK & pK & K & K & K \\
\cdots & & p^2K & p^2K & p^2K & pK & pK & pK & K & K & K \\
\cdots & & & p^2K & p^2K & p^2K & pK & pK & pK & K & K & K
\end{array}
$$

$$\underbrace{}_{n} \qquad \underbrace{}_{n} \qquad \underbrace{}_{n}$$

Let G be a Sylow p-subgroup of $\mathbf{GL}_n(K)$. We know from Exercise 11.3.3, Ch. 4, that G consists of all $n \times n$ matrices over K congruent to the identity matrix modulo the matrix of ideals formed with the carpet in the indicated position. It turns out that for $r \ge 2$ the subgroup $\gamma_r G$ consists precisely of those matrices in $\mathbf{SL}_n(K)$ that are congruent to the identity modulo the matrix formed when the carpet of ideals is moved $r-1$ steps to the right. Let us now translate this into more formal language. Let K be an associative ring with a multiplicative identity. A family $\mathfrak{a} = \{\mathfrak{a}_{ij} \mid i, j \in \mathbf{Z}\}$ of ideals of K is called a *carpet of ideals* if

$$\mathfrak{a}_{ik}\mathfrak{a}_{kj} \subseteq \mathfrak{a}_{ij} \quad \text{for all } i, j, k \in \mathbf{Z}.$$

It is easily verified that if K is commutative then the set

$$\Gamma_n(\mathfrak{a}) = \{x \mid x \in \mathbf{SL}_n(K), x_{ij} \equiv \delta_{ij} \bmod \mathfrak{a}_{ij}\}$$

is a group; we shall call it the *(special) congruence subgroup modulo the carpet* $\mathfrak{a}$. (Note that if $K = \mathbf{Z}$, $\mathfrak{a}_{ij} = (m)$, this becomes the principal congruence subgroup $\Gamma_n(m)$, of Exercise 4.2.7.) In this notation our result can be restated (in terms of formulae rather than diagrams) as follows: If G is a Sylow p-subgroup of $\mathbf{GL}_n(K)$; $K = \mathbf{Z}_{p^m}$, then

$$\gamma_r G = \Gamma_n(\mathfrak{a}^{(n,r)}), \qquad r = 2, 3, \ldots,$$

where

$$\mathfrak{a}_{ij}^{(n,r)} = p^{-[(j-i-r)/n]}K;$$

here "[ ]" means "the integral part of," and for $l \leq 0$, we define $p^l K = K$. From this result it follows in particular that a Sylow p-subgroup of $\mathbf{GL}_n(\mathbf{Z}_{p^m})$ is nilpotent of class $mn - 1$.

It is clear from the statement following (4), that a group G is nilpotent of class $\leq s$ if and only if $\gamma_{s+1} G = 1$. Hence the nilpotent groups of class $\leq s$ form a variety, denoted by $\mathfrak{N}_s$; this variety is defined by the single law

$$[x_1, \ldots, x_{s+1}] = e$$

(see Exercise 14.4.1). Thus subgroups, homomorphic images, and Cartesian products of nilpotent groups of class $\leq s$, are again nilpotent of class $\leq s$. By the general theory set forth in Chapter 5, the variety $\mathfrak{N}_s$ possesses free groups; these are the groups

$$\mathbf{F}(X)/\gamma_{s+1}\mathbf{F}(X),$$

for any alphabet X.

16.1.3. Exercise. Let G be the free nilpotent group of class 2 with free generators a, b. Put $c = [a, b]$. Each $g \in G$ can be uniquely expressed in the form $g = a^\alpha b^\beta c^\gamma$, for integral α, β, γ; these expressions multiply as follows:

$$a^\alpha b^\beta c^\gamma \cdot a^{\alpha'} b^{\beta'} c^{\gamma'} = a^{\alpha+\alpha'} b^{\beta+\beta'} c^{\gamma+\gamma'+\alpha'\beta}.$$

Hence the map defined by

$$a^\alpha b^\beta c^\gamma \to \begin{pmatrix} 1 & \beta & \gamma \\ 0 & 1 & \alpha \\ 0 & 0 & 1 \end{pmatrix},$$

is an isomorphism between G and $\mathbf{UT}_3(\mathbf{Z})$.

We end this subsection with a few words about terminology. Nilpotent groups were not so named right from their inception: for a long time they went under the nondescript title of "special." In ring theory however the concept of nilpotency (i.e. "nil potency" or "zero power") had long been in use: a nilpotent ring is one in which, for some positive integer n, the product of any n elements is zero; i.e.

$$x_1 \cdots x_n = 0 \tag{5}$$

is a law in the ring (where we allow all bracketings if the ring is non-associative). (As an example take the ring of upper triangular matrices of degree n with zero main diagonal.) In the classical theory of Lie there is an important correspondence between a special class of rings (Lie rings), and a special class of groups (Lie groups), which matches multiplication in the rings with commutation (i.e. formation of commutators) in the groups; in particular, corresponding to the law (5) we get the law

$$[x_1, \ldots, x_n] = e. \tag{6}$$

For this reason "nilpotent" ultimately became the accepted name for groups satisfying the law (6). We note that group theory did not long remain indebted to ring theory: soluble Lie rings take their name from soluble groups.

The following exercise demonstrates another way of relating nilpotency in groups to nilpotency in rings.

16.1.4. Exercise. Let A be an associative ring with 1, B a subring, and write B^n for the subring generated by all products $b_1 \cdots b_n$, $b_i \in B$. If B is nilpotent, i.e. $B^n = 0$ for some n, then the set $G^{(i)} = \{1 + x \mid x \in B^i\}$ is a group under the ring multiplication, and $[G^{(i)}, G^{(j)}] \leq G^{(i+j)}$. It follows that these groups are all nilpotent of class $\leq n$. This gives another way of establishing the nilpotency of some of the groups mentioned in Example 16.1.2 above, without, however, obtaining a description of their upper or lower central series.

16.1.5. Exercise. In a torsion-free nilpotent group the identity element is the only element conjugate with its inverse.

16.1.6. Exercise. A periodic, finitely generated, nilpotent group is finite. (This is still true with "soluble" replacing "nilpotent.") (Hint. Use Exercise 14.3.2.)

16.2. General Properties

The very definition of nilpotent groups suggests an effective, and usually indispensible, tool for investigating them: induction on the nilpotency class. Although inductive proofs of this sort usually involve the upper or lower central series, on occasion it is convenient to work with other subgroups. In this connexion the following lemma is useful.

16.2.1. Lemma. *Let G be a nilpotent group of class $c \geq 2$. Any subgroup of G generated by the commutator subgroup together with any single element of G, is nilpotent of class $<c$.*

PROOF. Let $a \in G$ and write $H = \langle a, [G, G] \rangle$. Since

$$[G, G] \leq (\zeta_{c-1} G) \cap H \leq \zeta_{c-1} H,$$

it follows that $H/\zeta_{c-1} H$ is cyclic. Since in any group the factor group by the centre cannot be nontrivial cyclic, we must have $\zeta_{c-1} H = H$, as required.

This subsection is devoted mainly to establishing facts about subgroups of nilpotent groups.

16.2.2. Theorem. *Every subgroup of a nilpotent group is subnormal. More precisely, if G is nilpotent of class c then for any subgroup H the series of successive normalizers starting with H reaches G after at most c steps.*

PROOF. Write

$$Z_i = \zeta_i G, \qquad H_0 = H, \qquad H_{j+1} = N_G(H_j).$$

The theorem will follow if we can show that $Z_i \leq H_i$. For $i = 0$ this is obvious. We shall show that if it is true for i then it is true for $i + 1$. Since

$$[G, Z_{i+1}] \leq Z_i \leq H_i,$$

we have that

$$H_i^{Z_{i+1}} \subseteq H_i[H_i, Z_{i+1}] \leq H_i.$$

(Recall that for $A, B \leq G$, $A^B = \{a^b \mid a \in A, b \in B\}$.) Thus Z_{i+1} normalizes H_i; i.e. $Z_{i+1} \leq H_{i+1}$, which completes the inductive step and the proof.

16.2.3. Theorem. *In a nilpotent group every nontrivial normal subgroup intersects the center nontrivially.*

PROOF. We use induction on the nilpotency class. Let G be a nilpotent group and H a nontrivial normal subgroup. Suppose inductively that the theorem is true for groups of smaller class than G. As usual write $Z_i = \zeta_i G$. If $H \leq Z_1$ we have the desired conclusion trivially. Suppose $H \not\leq Z_1$. Then by the inductive hypothesis applied to G/Z_1, the intersection $HZ_1 \cap Z_2$ contains an element $a \notin Z_1$. If we write $a = hz$ for some $h \in H$, $z \in Z_1$, then $h \in H \cap Z_2$, $h \notin Z_1$. Let $g \in G$ be any element such that $[h, g] \neq e$. Then

$$[h, g] \in H \cap [Z_2, G] \leq H \cap Z_1,$$

so that the intersection $H \cap Z_1$ is nontrivial, and the theorem is proved.

16.2.4. Exercise. In a nilpotent group every normal subgroup of prime order is contained in the center.

16.2.5. Theorem. *Let G be a nilpotent group. If A is a subgroup such that $A[G, G] = G$, then $A = G$. Hence*

$$[G, G] \leq \Phi(G). \tag{7}$$

PROOF. Suppose on the contrary that $A \neq G$. Write $A_i = A \cdot \zeta_i G$, $i = 0, 1, 2, \ldots$; clearly $A_i \trianglelefteq A_{i+1}$. Suppose n is such that $A_n < G$, $A_{n+1} = G$. Since the quotient A_{n+1}/A_n is abelian we have $[G, G] \leq A_n$, whence

$$A[G, G] \leq A_n < G,$$

a contradiction. The assertion (7) about the Frattini subgroup follows directly from its description as the set of nongenerators of G (Theorem 2.2.6). This completes the proof.

[At this point we reveal a secret. In our fourth example on the Frattini subgroup in 2.2.7, we proved that $\Phi(\mathbf{UT}_n(\mathbf{Z})) \leq \mathbf{UT}_n^2(\mathbf{Z})$, and mentioned (leaving the onus of proof on the reader!) that the reverse inclusion is also true. What we had in mind was precisely the theorem just proved.]

16.2.6. Theorem. *In a nilpotent group G every maximal abelian normal subgroup A is its own centralizer. It follows that A is a maximal abelian subgroup and G/A embeds in* Aut A.

PROOF. Write briefly $H = C_G(A)$, $Z_i = \zeta_i(G)$. Suppose inductively that $H \cap Z_i \leq A$ (this being trivial for $i = 0$), and let $x \in H \cap Z_{i+1}$. For all $g \in G$, we have $[x, g] \in H \cap Z_i \leq A$; hence $\langle x, A \rangle$ is an abelian normal subgroup of G, containing A. By the maximality of A, we must therefore have that $x \in A$, whence $H \cap Z_{i+1} \leq A$. Since $Z_n = G$ for some n, we get that $H = A$, as we wished to prove.

We shall call elements of finite order in a group *torsion elements*.

16.2.7. Theorem. *In a nilpotent group G the set τG of torsion elements forms a subgroup* (the *torsion subgroup* of G).

PROOF. We shall use induction on the nilpotency class of G, together with Lemma 16.2.1. The result is easy if G has class 1; suppose it has class >1 and that the theorem is true for groups of smaller class. Let a, b be arbitrary torsion elements of the group G. Set

$$A = \langle a, [G, G] \rangle, \qquad B = \langle b, [G, G] \rangle.$$

By the inductive hypothesis τA, τB are subgroups of A, B respectively. Since τA is fully invariant in A, and A is normal in G, it follows that τA is normal in G. Similarly τB is normal in G. It is easy to see that for each element of $\tau A \cdot \tau B$ there is some power of that element which lies in τB, and therefore some higher power which is trivial. Hence $\tau A \cdot \tau B$ is periodic. In particular the elements ab, a^{-1} have finite order, as we were required to prove.

16.2.8. Theorem. *In a torsion-free nilpotent group roots of elements, when they exist, are unique; i.e. if $a^n = b^n$, $n \neq 0$, then $a = b$.*

PROOF. Yet again we use induction on the nilpotency of our torsion-free group G, say. For G abelian the result is obvious; suppose G is not abelian and that the result holds for torsion-free groups of smaller class. The subgroup $\langle a, [G, G] \rangle$ is normal in G and has smaller class (Lemma 16.2.1). Since $a, a^b \in \langle a, [G, G] \rangle$, and, as is easy to see, $(a^b)^n = a^n$, it follows from the inductive hypothesis that $a^b = a$, that is, a commutes with b. Hence $(ab^{-1})^n = e$. Since G is torsion-free we deduce that $a = b$, as desired.

16.2.9. Exercise. In a torsion-free nilpotent group if $x^m y^n = y^n x^m$ ($m, n \neq 0$) then $xy = yx$. (Hint. If $(y^{-n} x y^n)^m = x^m$, then $y^{-n} x y^n = x$, etc.)

16.2.10. Exercise. The factors of the upper central series of a torsion-free nilpotent group are also torsion-free.

16.2.11. Exercise. The analogous statement for the lower central series is false: for example, if

$$G = \begin{vmatrix} 1 & n\mathbf{Z} & \mathbf{Z} \\ 0 & 1 & \mathbf{Z} \\ 0 & 0 & 1 \end{vmatrix}, \quad \text{then } [G, G] = \begin{vmatrix} 1 & 0 & n\mathbf{Z} \\ 0 & 1 & 0 \\ 0 & 0 & 1 \end{vmatrix},$$

whence $G/[G, G] \simeq \mathbf{Z} \oplus \mathbf{Z} \oplus \mathbf{Z}_n$.

We shall now consider nilpotent normal subgroups of arbitrary groups.

16.2.12. Theorem (Fitting). *In any group the product of two nilpotent normal subgroups of classes s, t is a nilpotent normal subgroup of class $\leq s + t$.*

PROOF. Let $A \trianglelefteq G$, $B \trianglelefteq G$, $\gamma_{s+1} A = 1$, $\gamma_{t+1} B = 1$. Then

$$\gamma_n(AB) = [\underbrace{AB, \ldots, AB}_{n}] = \prod [H_1, \ldots, H_n], \tag{8}$$

where $(H_1, \ldots, H_n)$ ranges over all sequences of n A's and B's (see Exercise 3.2.11). Since $\gamma_i A$ is fully invariant in A, which in turn is normal in G, it follows that $\gamma_i A$ is normal in G, whence

$$[\gamma_i A, B] \leq \gamma_i A \quad \text{for all } i.$$

It follows that if i of the entries $H_1, \ldots, H_n$ are A's, and the remaining $n - i$ are B's, then

$$[H_1, \ldots, H_n] \leq \gamma_i A \cap \gamma_{n-i} B. \tag{9}$$

If we take $n = s + t + 1$, then either $i \geq s + 1$ or $n - i \geq t + 1$, whence by (8) and (9) $\gamma_n(AB) = 1$. The proof is complete.

16.2.13. Exercise. Define a group G to be *residually nilpotent* if

$$\gamma_\omega G \equiv \bigcap_{n=1}^{\infty} \gamma_n G = 1.$$

The product of two normal residually nilpotent subgroups of a group need not be residually nilpotent: a counterexample is provided by the equality

$$\mathbf{SL}_n(\mathbf{Z}) = \Gamma_n(2) \cdot \Gamma_n(3).$$

16.3. Nilpotent Groups of Automorphisms

As noted in Example 16.1.2, any group of unitriangular matrices is nilpotent. Now of course matrices of given size over a given field are (or at least arise from) automorphisms of a vector space with a given basis, and from this point of view the unitriangular matrices are just those automorphisms fixing each member of a descending chain of subspaces spanned by subsets of the basis, and acting trivially on the factors of the chain. Herein lies the real reason for the nilpotency of groups of unitriangular matrices. More generally we have

16.3.1. Theorem (Kalužnin). *Suppose that the series*

$$G = G_0 \geq G_1 \geq \cdots \geq G_r = 1 \tag{10}$$

is a normal series (of length r) of the group G. Let Φ denote the stabilizer of the series (10), i.e. the group of all automorphisms of G leaving the G_i invariant and acting trivially on the factors G_i/G_{i+1}. Then Φ is nilpotent of class $<r$, and if we regard G, Φ as subgroups of $\mathrm{Hol}\, G$, then $[G, \Phi]$ is also nilpotent of class $<r$.

PROOF. To see that Φ is nilpotent of class $<r$, note first that by the hypothesis

$$[G_i, \Phi] \leq G_{i+1} \quad \text{for all } i. \tag{11}$$

Write Φ_j for the set of elements of Φ acting trivially on *all* factors G_i/G_{i+j}, $i = 0, 1, \ldots, r - j$. It is obvious that

$$\Phi = \Phi_1 \geq \Phi_2 \geq \cdots \geq \Phi_r = 1,$$

and it suffices to show that this is a central series for Φ; i.e. that

$$[\Phi_j, \Phi] \leq \Phi_{j+1} \quad \text{for all } j,$$

or, equivalently, that

$$[G_i, [\Phi_j, \Phi]] \leq G_{i+j+1} \quad \text{for all appropriate } i, j.$$

But this is immediate from the three-commutator lemma (3.2.10) and the following easy inclusions:

$$[\Phi_j, [\Phi, G_i]] \leq [\Phi_j, G_{i+1}] \leq G_{i+j+1};$$

$$[\Phi, [G_i, \Phi_j]] \leq [\Phi, G_{i+j}] \leq G_{i+j+1}.$$

(ii) We wish to prove the same thing for $[G, \Phi]$. From the inclusions

$$[G, [\Phi, G_i]] \leq [G, G_{i+1}] \leq G_{i+1},$$

$$[\Phi, [G_i, G]] \leq [\Phi, G_i] \leq G_{i+1},$$

and, once again, the lemma on the three commutators, we deduce that

$$[G_i, [G, \Phi]] \leq G_{i+1} \quad \text{for all } i.$$

This means that the actions of $[G, \Phi]$ on the factors of the series $G_1 \geq G_2 \geq \cdots \geq G_r = 1$ by conjugation, are trivial. Hence under conjugation $[G, \Phi]$ stabilizes this series (we are also using here the normality of the series (10)). Thus by Part (i) of the proof $[G, \Phi]/C_{[G,\Phi]}(G_1)$ is nilpotent of class $< r - 1$. Since $[G, \Phi] \leq G_1$, we have that $C_{[G,\Phi]}(G_1)$ lies in the center of $[G, \Phi]$, whence the desired conclusion.

It is natural to try to go further: Is it perhaps true that the stabilizer of an *arbitrary* series (not necessarily normal) of subgroups is nilpotent? (By *stabilizer of a (not necessarily normal) series* (10) we mean here the group of all elements of Aut G leaving the G_i invariant and satisfying (11).) It turns out that the answer is affirmative, although the proof is more complicated, and the bound on the nilpotency class is not as good.

16.3.2. Theorem (P. Hall). *The stabilizer of series of subgroups of length r is a nilpotent group of class* $\leq \binom{r}{2}$.

PROOF. We use induction on the length r of the series. The result is trivial for $r = 1$. Suppose Φ to be the stabilizer of the series (10) where we assume $r > 1$. Put $\Psi = C_\Phi(G_1)$. Since Φ stabilizes the series $G_1 \geq G_2 \geq \cdots \geq G_r$ of length $r - 1$, by the (tacit) inductive hypothesis the group Φ/Ψ is nilpotent of class $\leq \binom{r-1}{2}$; i.e.

$$\gamma_{1+\binom{r-1}{2}}\Phi \leq \Psi.$$

Put

$$\Psi_1 = \Psi, \qquad \Psi_{i+1} = [\Psi_i, \Phi].$$

Since $\Psi \trianglelefteq \Phi$, we have

$$\Psi = \Psi_1 \geq \Psi_2 \geq \cdots.$$

Thus it suffices to show that $[G, \Psi_r] = 1$. We do this by proving by induction on i that $[G, \Psi_i] \le G_i$ for all i. This is immediate for $i = 1$. For the inductive step, we take $g \in G$, $\psi_{i+1} \in \Psi_{i+1}$, and prove that $[\psi_{i+1}, g] \in G_{i+1}$. Now ψ_{i+1} is a word in commutators of the form

$$[\psi_i, \phi^{-1}], \qquad \psi_i \in \Psi_i, \qquad \phi \in \Phi.$$

The element ψ_{i+1}^g is the *same* word in the elements

$$[\psi_i, \phi^{-1}]^g = [\psi_i, \phi^{-1}][\psi_i, \phi^{-1}, g]. \tag{12}$$

From Witt's identity (see (17) of §3, Ch. 1) and the containments

$$[\phi, g^{-1}, \psi_i] \in [\Phi, G, \Psi_i] \le [G_1, \Psi] = 1,$$

$$[g, \psi_i^{-1}, \phi] \in [G, \Psi_i, \Phi] \le [G_i, \Phi] \le G_{i+1}$$

(where we have invoked the inductive hypothesis), we get

$$[\psi_i, \phi^{-1}, g] \in G_{i+1}.$$

Hence in the product in (12) the left-hand factor lies in Ψ, and the right-hand factor in G_{i+1}. But Ψ centralizes $G_1 \ge G_{i+1}$, whence $\psi_{i+1}^g \in \psi_{i+1} G_{i+1}$. Thus $[\psi_{i+1}, g] \in G_{i+1}$, completing both inductive steps, and thereby the proof.

Our demure Example 16.1.2 reveals unsuspected depth!

§17. The Most Important Subclasses

17.1. Finite Nilpotent Groups

In the following example we introduce another important source of nilpotent groups.

17.1.1. EXAMPLE. *Every finite p-group G is nilpotent.* The bulk of the proof of this consists in showing that a p-group has nontrivial center. Observe first that an element is central if and only if it is the only member of its conjugacy class. Since

$$|a^G| = |G : N_G(a)|,$$

the sizes c_i of the conjugacy classes in G are powers of p. Since the identity element forms a conjugacy class by itself, at least one of the c_i is 1. However $\sum c_i = |G|$, so that p divides $\sum c_i$; this and the fact that $p \mid c_i$ for $c_i > 1$ imply that at least one (in fact at least $p - 1$) other $c_i = 1$. Hence G has nontrivial center Z_1. The nilpotency of G now follows easily, since for the same reason G/Z_1 has nontrivial center Z_2/Z_1, and so on, until finally one of the higher centers coincides with G.

17.1.2. Exercise. For each prime p there are exactly two non-isomorphic groups of order p^2, namely $\mathbf{Z}_{p^2}$ and $\mathbf{Z}_p \oplus \mathbf{Z}_p$. There are five of order p^3. Of these three are abelian:

$$\mathbf{Z}_{p^3}, \mathbf{Z}_{p^2} \oplus \mathbf{Z}_p, \mathbf{Z}_p \oplus \mathbf{Z}_p \oplus \mathbf{Z}_p;$$

and two are nonabelian: for $p = 2$ these are the *dihedral group*

$$\langle a, b \,|\, a^4 = 1, b^2 = 1, bab = a^{-1} \rangle,$$

and the *quaternion group*

$$\langle a, b \,|\, a^4 = 1, b^2 = a^2, b^{-1}ab = a^{-1} \rangle;$$

while for $p > 2$ they are the groups

$$\langle a, b \,|\, a^{p^2} = 1, b^p = 1, b^{-1}ab = a^{1+p} \rangle,$$

$$\langle a, b, c \,|\, a^p = 1, b^p = 1, c^p = 1, [a, b] = c, [a, c] = [b, c] = 1 \rangle.$$

An infinite p-group may have trivial center. By Exercise 6.2.3 the restricted wreath product of an arbitrary group with an infinite group has trivial center: hence the p-group $\mathbf{C}_p \text{ wr } \mathbf{C}_{p^\infty}$ will serve as an example.

17.1.3. Exercise. Let K be a field of non-zero characteristic p, and denote by $\mathbf{UT}_\omega(K)$ the group of those infinite matrices with rows and columns indexed by the natural numbers, entries on the main diagonal all 1, entries below the main diagonal all 0, and above the main diagonal only finitely many non-zero entries. The group $\mathbf{UT}_\omega(K)$ is a p-group with trivial center. (Hint. Use (2) of §3, Ch. 1.)

It turns out that a finite nilpotent group is nothing more than a direct product of finitely many finite p-groups (for various p of course). We include this pleasant fact as part of

17.1.4. Theorem (Burnside–Wielandt). *Let G be a finite group. The following conditions are equivalent:*

   (*i*) *G is nilpotent;*
  (*ii*) *every subgroup of G is subnormal;*
 (*iii*) *G is the direct product of its Sylow p-subgroups;*
 (*iv*) *$[G, G] \le \Phi(G)$.*

PROOF. We shall prove the theorem according to the following scheme:

(i) $\Rightarrow$ (ii) See Theorem 16.2.2.

(ii) $\Rightarrow$ (iii) Let P be a Sylow p-subgroup of G. Since $N_G(P)$ is its own normalizer in G (Exercise 11.1.4), while by (ii) every proper subgroup has normalizer strictly larger than itself, we must have that $N_G(P) = G$. Hence the Sylow subgroups of G are normal in G. Statement (iii) now follows easily.

(iii) $\Rightarrow$ (i) Use Example 17.1.1.

(i) $\Rightarrow$ (iv) See Theorem 16.2.5.

(iv) $\Rightarrow$ (iii) As before it suffices to show that the Sylow subgroups of G are normal in G. Let P be a Sylow p-subgroup of G and suppose on the contrary that $N_G(P)$ is proper in G. Let H be a maximal subgroup of G containing $N_G(P)$. Since

$$[G, G] \le \Phi[G] \le H,$$

it follows that H is normal in G. On the other hand since H contains $N_G(P)$, it is its own normalizer in G (Exercise 11.1.4). This contradiction completes the proof of the theorem.

From the Burnside–Wielandt theorem it follows immediately that the wreath product of a finite nontrivial p-group with a finite nontrivial q-group is nilpotent if and only if $p = q$.

17.1.5. Exercise. What is the nilpotency class of the wreath product $\mathbf{Z}_{p^n}$ wr $\mathbf{Z}_p$?

The precise conditions for wreath products to be nilpotent are given in the next exercise.

17.1.6. Exercise. Let A, B be nontrivial nilpotent groups. The wreath products A wr B, A Wr B are each nilpotent if and only if A and B are p-groups (for the same p), with A of finite exponent and B finite. (Hint. Use Exercises 6.2.1, 6.2.3, and the natural homomorphism $\mathbf{Z}$ wr $\mathbf{Z}_p \to \mathbf{Z}_q$ wr $\mathbf{Z}_p$.)

We conclude this subsection with the following striking fact about arbitrary finite groups.

17.1.7. Theorem (Frattini). *The Frattini subgroup of a finite group is nilpotent.*

PROOF. Let G be a finite group and A its Frattini subgroup. In view of the Burnside–Wielandt theorem it suffices to show that the Sylow subgroups of A are normal in A, or, equivalently, in G (see Exercise 5.2.6); i.e. that for any Sylow p-subgroup P of A, $N_G(P) = G$. But this is equivalent to showing that

$$G = A \cdot N_G(P)$$

since the set A is finite and consists of the nongenerators of the group G (Theorem 2.2.6). We shall now establish the latter equality.

Let g be any element of G. Conjugation of G by g sends A to itself, whence P^g is also a Sylow p-subgroup of A. By Sylow's theorem (11.1.1) P and P^g are conjugate in A, i.e. $P^g = P^a$ for some $a \in A$. Hence $ga^{-1} \in N_G(P)$, and finally $g \in A \cdot N_G(P)$.

A part of this proof is often isolated as the "Frattini argument," the object of which is

17.1.8. The Frattini Lemma. *Let A be a normal subgroup of a finite group G, and let P be a Sylow subgroup of A. Then $G = A \cdot N_G(P)$.*

17.1.9. Exercise. Let A be a normal subgroup of a group G, and let B be a subgroup of A with the property that whenever $B_1 \le A$ is conjugate to B in G then it is conjugate to B in A. Then $G = A \cdot N_G(B)$. (This is the "Generalized Frattini Lemma.")

17.2. Finitely Generated Nilpotent Groups

The groups $\mathbf{UT}_n(\mathbf{Z})$ are obvious examples. It is not obvious, but it is true, that the subgroups of the unitriangular groups over $\mathbf{Z}$ account (up to isomorphism of course) for all finitely generated, torsion-free nilpotent groups. Arbitrary finitely generated nilpotent groups are just finite extensions of the torsion-free ones, and so, in particular, can be embedded in $\mathbf{SL}_n(\mathbf{Z})$ for suitable n. The goal of this subsection will be to prove these facts, fundamental in the theory of finitely generated nilpotent groups.

17.2.1. Lemma. *Let G be an arbitrary group, and let M be a set of generators for G. Then $\gamma_i G$ is generated by the elements of $\gamma_{i+1} G$ together with all left-normed commutators of weight i in the elements of M.*

PROOF. The statement is obvious for $i = 1$. We go to the inductive step from i to $i+1$. By definition $\gamma_{i+1} G$ is generated by the elements $[x, y]$, $x \in \gamma_i G$, $y \in G$. By the inductive hypothesis $x = x_1^{\varepsilon_1} \cdots x_n^{\varepsilon_n} z$, where each x_j is a left-normed commutator of weight i in the elements of M, $\varepsilon_j = \pm 1$, and $z \in \gamma_{i+1} G$. Expressing y as a product of elements of $M \cup M^{-1}$, and using the commutator identities (3) of §3.2, Ch. 1, we see that $[x, y]$ is a word in elements of the form

$$[x_j, a]^g = [x_j, a][x_j, a, g], \qquad [z, a]^g, \qquad a \in M, g \in G.$$

Since $[x_j, a, g]$ and $[z, a]$ belong to $\gamma_{i+2} G$, and $[x_j, a]$ is a left-normed commutator of weight $i+1$ in the elements of M, the proof is complete.

We shall say that a group *almost* possesses a certain property if it has a normal subgroup of finite index with the property.

17.2.2. Theorem. *Every finitely generated nilpotent group has a central series with cyclic factors, and is almost torsion-free. A torsion-free, finitely generated nilpotent group has a central series with infinite cyclic factors.*

PROOF. (i) Let G be finitely generated nilpotent. Each factor of its lower central series is a finitely generated abelian group (by the preceding lemma), and therefore possesses a (finite) series with cyclic factors. Since any refinement of a central series is again a central series, it follows that G has a central series with cyclic factors. We now use induction on the length of this central series to prove that G is almost torsion-free. This is trivial if the series has length 1; suppose it has length >1 and that groups with shorter such central series are almost torsion-free. Let H be the largest member of the series different from G, and let a be an element generating G modulo H. By the inductive hypothesis H is almost torsion-free, so that for a suitable n its subgroup $H^n = \langle h^n \mid h \in H \rangle$ is torsion-free. By Exercise 16.1.6 we have $|H:H^n| < \infty$. If $|G:H| < \infty$, then H^n will serve as the desired normal torsion-free subgroup of finite index in G. If $|G:H| = \infty$, then the desired subgroup may be taken to be the intersection of the conjugates of the subgroup $\langle a \rangle \cdot H^n$, since this subgroup is torsion-free and

$$|G: \langle a \rangle H^n| = |H:H \cap \langle a \rangle H^n| = |H:H^n| < \infty.$$

(ii) By Part (i) any finitely generated nilpotent group is polycyclic, whence the factors of its upper central series are also polycyclic (Exercise 4.4.3), and therefore finitely generated. If our nilpotent group is torsion-free then they also will be torsion-free (Exercise 16.2.10). Hence the upper central series of a finitely generated torsion-free nilpotent group can be refined to a central series with infinite cyclic factors. This completes the proof.

17.2.3. Exercise. Polycyclic groups are almost torsion-free. (Hint. Imitate part of the proof of the preceding theorem.)

17.2.4. Exercise. The torsion subgroup of a finitely generated nilpotent group is finite. More generally, in any group with almost no torsion all periodic subgroups are finite.

From Theorem 17.2.2 it follows that a finitely generated torsion-free nilpotent group G can be equipped with a system of integral coordinates of a certain special kind, which lead to a representation of G by integral unitriangular matrices. We now give the details of this. For the time being take G to be simply a set. A (finite) sequence of functions $f_i: G \to \mathbf{Z}$, $i = 1, \ldots, s$, is called a *coordinate system* for G if the map defined by $x \to (f_1(x), \ldots, f_s(x))$ is an injection from G to the set $\mathbf{Z}^s$ of all ordered

s-tuples of integers. For the next definition we suppose that $G \subseteq \mathbf{Z}^s$. A map $\phi: G \to \mathbf{Z}^r$ is a *polynomial map* if there exist polynomials $f_1, \ldots, f_r$ over $\mathbf{Q}$, in s variables, such that

$$x\phi = (f_1(x), \ldots, f_r(x)) \quad \text{for all } x \in G.$$

If $f_1, \ldots, f_r$ have degree one, then we shall say that ϕ is *linear*. Now let G be a finitely generated torsion-free nilpotent group, and let

$$G = G_1 > G_2 > \cdots > G_{s+1} = 1$$

be a central series for G with infinite cyclic factors. Choose elements $a_1, \cdots, a_s$ such that $G_i = \langle a_i, G_{i+1} \rangle$. Obviously each element x of G can be uniquely expressed in the form

$$x = a_1^{t_1(x)} \cdots a_s^{t_s(x)}, \qquad t_i(x) \in \mathbf{Z},$$

so that the s-tuple of functions $t_1, \ldots, t_s$ is a coordinate system for G. We call the s-tuple $(a_1, \ldots, a_s)$ a *Mal'cev basis* (of order s) for the group G, and the integers $t_1(x), \ldots, t_s(x)$ the *Mal'cev coordinates* of x relative to this basis.

17.2.5. Theorem. *Let G be a finitely generated torsion-free nilpotent group and $(t_1, \ldots, t_s)$ a Mal'cev coordinate system for G. There exists a positive integer $n = n(G)$ and a monomorphism $\phi: G \to \mathbf{UT}_n(\mathbf{Z})$, such that ϕ is a polynomial map on G, and its inverse ϕ^{-1} is linear on G^ϕ (here we are identifying G with a subset of $\mathbf{Z}^s$ according to the given coordinate system, and $\mathbf{UT}_n(\mathbf{Z})$ with a subset of $\mathbf{Z}^{n^2}$ in the obvious way). It follows in particular that the operations of multiplication and powering in the group G can be described in terms of polynomials in the Mal'cev coordinates. More precisely for $x, y \in G$, $m \in \mathbf{Z}$, $1 \le i \le s$, we have*

$$t_i(xy) = [a \text{ polynomial over } \mathbf{Q} \text{ in } \{t_\alpha(x), t_\alpha(y) \mid \alpha < i\}] + t_i(x) + t_i(y), \quad (1)$$

$$t_i(x^m) = [a \text{ polynomial over } \mathbf{Q} \text{ in } m \text{ and } \{t_\alpha(x) \mid \alpha < i\}] + m t_i(x). \quad (2)$$

PROOF. (i) We first prove the last statement: "It follows in particular ...." For any matrix $a \in \mathbf{UT}_n(\mathbf{Z})$ and any integer m we have by the binomial theorem

$$a^m = \sum_{i=0}^{n-1} \binom{m}{i} (a - e)^i, \quad \text{where of course } \binom{m}{i}$$

$$= \frac{m(m-1) \cdots (m-i+1)}{i!}, \quad \binom{m}{0} = 1. \quad (3)$$

It follows that the entries of the matrix a^m are polynomials in m and the entries of a. By the main assertion of the theorem (which we have yet to prove) the coordinates $t_i(xy)$, $t_i(x^m)$ are linear polynomials in the entries of the matrices $(xy)^\phi$, $(x^m)^\phi$ respectively, which entries are, as we have just seen, polynomials in the integer m and the entries of the matrices x^ϕ, y^ϕ. Again by the main assertion of the theorem the latter entries are polynomials in the coordinates $t_\alpha(x)$, $t_\alpha(y)$. Hence the Mal'cev coordinates of the product xy and the power x^m are polynomials in the integer m and the Mal'cev coordinates of x and y, as required. That these polynomials are as described in (1) and (2) follows easily from the definition of a central series and Mal'cev coordinates.

(ii) We now prove the main assertion. First note that instead of ϕ it suffices to find a monomorphism $\psi: G \to \mathbf{GL}_n(\mathbf{Z})$ which is polynomial, and whose inverse ψ^{-1} is linear on G^ψ, such that in addition G^ψ consists of unipotent matrices. (A matrix a is said to be *unipotent* if its only characteristic root is 1, or, equivalently, if $(a - e)^n = 0$.) This is immediate from the fact that such a group G^ψ is conjugate in $\mathbf{GL}_n(\mathbf{Q})$ to a subgroup of $\mathbf{UT}_n(\mathbf{Z})$. We shall now prove this for any nilpotent group H of unipotent matrices over $\mathbf{Z}$. The first step is to show that H is conjugate to a subgroup of $\mathbf{UT}_n(\mathbf{Q})$. Regarding $\mathbf{GL}_n(\mathbf{Q})$ as the automorphism group of an n-dimensional vector space V over the field $\mathbf{Q}$, let

$$V = V_1 > V_2 > \cdots > V_{m+1} = 0 \tag{4}$$

be an unrefinable chain of subspaces of V invariant under H. Choose a basis for V such that subsets of it span the V_i, $i = 1, \ldots, m + 1$; relative to such a basis each automorphism in H can be written as a block matrix whose blocks below the main diagonal are all zero, and whose ith diagonal block arises from the action of H on V_i/V_{i+1}. We shall show that the action of H on each V_i/V_{i+1} is trivial. Let U be any V_i/V_{i+1}. Since $[H, H]$ has smaller nilpotency class than H, we may suppose (inductively), that (4) refines to a chain of subspaces invariant under $[H, H]$ whose factors have dimension 1, and are such that $[H, H]$ acts trivially on them. Thus there is a nonzero vector $u \in U$ fixed by every automorphism from $[H, H]$. Let $\hat{U}$ be the subspace of U consisting of all such vectors. This subspace is invariant under H since for $h \in H$, $h' \in [H, H]$ we have

$$(uh)h' = u(hh'h^{-1})h = uh.$$

Since U contains no proper subspaces invariant under H, we get that $\hat{U} = U$, whence $[H, H]$ acts trivially on U; i.e. H induces an abelian group of linear transformations of U. Now it is a fact of linear algebra that any set of pairwise commuting linear transformations over a field K have a common eigenvector over the extension field obtained by adjoining to the ground field K the characteristic roots of the linear transformations. It follows that the linear transformations of U induced by H, being unipotent, have a common eigenvector in U, which moreover they fix. Hence this vector spans a one-dimensional subspace of U on which H acts trivially. Since U contains no proper subspaces invariant under H, we deduce that U itself is a one-dimensional space on which H acts trivially. This shows that $H^a \leq \mathbf{UT}_n(\mathbf{Q})$ for some $a \in \mathbf{GL}_n(\mathbf{Q})$. Now take a finite set of matrices generating H^a, and denote by N a common denominator of their entries. If we write b for the diagonal matrix with entries $1, N, N^2, \ldots, N^{n-1}$ going down the main diagonal, then it is easy to see that $H^{ab} \leq \mathbf{UT}_n(\mathbf{Z})$.

(iii) It remains to find the map ψ. We shall use induction on the order of a Mal'cev basis. If this is zero, i.e. if the group is trivial, then the theorem is trivial. Thus suppose $s > 0$ and that for groups with a Mal'cev basis of order $< s$ the desired representation has been found; it follows from Parts (i) and (ii) of the proof that this amounts to assuming the theorem true for such groups. Suppose the group G has a Mal'cev basis $(a_1, \ldots, a_s)$ of length s. To prepare the ground for the construction of ψ we shall prove the statement (1) for G. Write briefly $t_i(x) = \xi_i$, $t_i(y) = \eta_i$. Obviously

$$xy = a_1^{\xi_1 + \eta_1}(a_1^{-\eta_1}a_2^{-1}a_1^{\eta_1})^{-\xi_2} \cdots (a_1^{-\eta_1}a_s^{-1}a_1^{\eta_1})^{-\xi_s}a_2^{\eta_2} \cdots a_s^{\eta_s},$$

and

$$a_1^{-\eta}a_i^{-1}a_1^{\eta} = [a_1^{\eta}, a_i]a_i^{-1}.$$

Since the theorem is true for groups with a Mal'cev basis of order $< s$ so, in particular, are (1) and (2) true for such groups; it will therefore suffice to verify that the coordinates of an element $[a_1^{\eta}, a_i]$ relative to the Mal'cev basis $(a_2, \ldots, a_s)$ are polynomials in η. Now

$$[a_1^{\eta}, a_i] = a_1^{-\eta}(a_i^{-1}a_1a_i)^{\eta},$$

and

$$a_i^{-1}a_1a_i = a_1a_{i+1}^{c_{i,i+1}} \cdots a_s^{c_{is}} \quad \text{for some } c_{ij} \in \mathbf{Z}.$$

From the inductive hypothesis as applied to $\langle a_1, a_{i+1}, \ldots, a_s \rangle$ we conclude that

$$(a_i^{-1}a_1a_i)^{\eta} = a_1^{\eta}a_{i+1}^{\zeta_{i,i+1}} \cdots a_s^{\zeta_{is}},$$

where the ζ_{ij} are polynomials in η. Hence

$$[a_1^{\eta}, a_i] = a_{i+1}^{\zeta_{i,i+1}} \cdots a_s^{\zeta_{is}},$$

and (1) is established for G.

(iv) We now construct ψ. Write $\mathbf{Q}[t_1, \ldots, t_s]$ for the ring of polynomials over $\mathbf{Q}$ in the functions $t_1, \ldots, t_s$ (where functions are multiplied and added "componentwise"), and define an action of G on this ring by setting, for each $a \in G$,

$$f^a(x) = f(ax), \qquad f \in \mathbf{Q}[t_1, \ldots, t_s], \qquad x \in G,$$

(i.e. by "left translation of the argument"). It is easy to see that this action of the element a defines an automorphism $\hat{a}$ say, of $\mathbf{Q}[t_1, \ldots, t_s]$, and that the map defined by $a \to \hat{a}$ defines a monomorphism $G \to \text{Aut}(\mathbf{Q}[t_1, \ldots, t_s])$.

An element of $\mathbf{Q}[t_1, \ldots, t_s]$ of the form $M(t_1, \ldots, t_s) = t_1^{m_1} \cdots t_s^{m_s}$ is called a *monomial* in $t_1, \ldots, t_s$. In view of (1) we have

$$t_i^a = t_i + \sum_j c_{ij}(a)M_j(t_1, \ldots, t_s), \tag{5}$$

where the $c_{ij}(a)$ are polynomials over $\mathbf{Q}$ in the Mal'cev coordinates of the element a, and the monomials $M_j(t_1, \ldots, t_s)$ occurring in the equation (5) with nonzero coefficients, do not involve $t_i, \ldots, t_s$. It follows that the ring endomorphism $\hat{a} - \hat{e}$ sends each monomial $M(t_1, \ldots, t_s)$ either to 0 or to a linear combination of lesser monomials, where we define $t_1^{m_1} \cdots t_s^{m_s} <$ $t_1^{n_1} \cdots t_s^{n_s}$ if for some k, $1 \le k \le s$, we have $m_i = n_i$ for $i > k$ but $m_k < n_k$. Thus for a sufficiently large $m = m(a, M)$ the endomorphism $(\hat{a} - \hat{e})^m$ annihilates $M(t_1, \ldots, t_s)$.

Let H be the additive subgroup generated by the images of the coordinate functions $t_1, \ldots, t_s$ under all $\hat{x}$, $x \in G$ (i.e. by the orbit of $\{t_1, \ldots, t_s\}$ under the action of G). We see from (5) that for some positive integer N, H is contained in the subgroup generated by the t_i, $i = 1, \ldots, s$ and the functions $(1/N)M_j(t_1, \ldots, t_s)$; hence this subgroup, and therefore also H, is finitely generated. Let $h_1, \ldots, h_n$ be a free basis for the free abelian group H. If we express h_k^x in terms of this basis:

$$h_k^x = \sum_l \psi_{kl}(x)h_l, \tag{6}$$

then $(\psi_{kl}(x))$ is the matrix of the restriction of $\hat{x}$ to H, relative to the basis $h_1, \ldots, h_n$. Since H contains the coordinate functions $t_1, \ldots, t_s$, the map $\psi: G \to \mathbf{GL}_n(\mathbf{Z})$ defined by $x \to (\psi_{kl}(x))$, is a monomorphism, and moreover G^ψ consists of unipotent matrices (since the characteristic polynomial of $\hat{x}|_H$ divides $(z - 1)^m$ for some m; see the preceding paragraph). The map ψ^{-1} is linear on G^ψ since the t_i are linear in the h_k, which are in turn linear in the ψ_{kl}; to see the latter statement evaluate both sides of (6) at e to get

$$h_k^x(e) = h_k(x) = \sum_l h_l(e)\psi_{kl}(x).$$

Finally we show that ψ is polynomial; i.e. that the functions ψ_{kl} are the restrictions to G (identified with $\mathbf{Z}^s$) of certain polynomials over $\mathbf{Q}$. Thus let $x \in G$. Since each h_k is a linear combination over $\mathbf{Z}$ of certain t_i^g, $g \in G$, it follows (using (5)) that h_k^x is the same linear combination of

$$t_i^{gx} = t_i + \sum_j c_{ij}(gx)M_j(t_1, \ldots, t_s).$$

Hence there are polynomials P_{kj} over $\mathbf{Q}$ such that

$$h_k^x = \sum_j P_{kj}(x)M_j(t_1, \ldots, t_s); \tag{7}$$

in particular

$$h_k = \sum_j P_{kj}(e)M_j(t_1, \ldots, t_s). \tag{8}$$

Since $h_1, \ldots, h_n$ are linearly independent over $\mathbf{Q}$, so are the rows of the matrix $(P_{kj}(e))$. Replacing h_k^x and h_l in (6) by the right-hand sides of (7) and (8), and using the linear independence of $\{M_j(t_1, \ldots, t_s)\}$, we obtain a system of linear equations in the $\psi_{kl}(x)$:

$$P_{kj}(x) = \sum_l \psi_{kl}(x)P_{lj}(e),$$

from which it is evident that the ψ_{kl} are polynomial over **Q**. This completes the proof of the theorem.

It is appropriate to mention in connexion with this proof that there is a general "method of split coordinates" [Ju. I. Merzljakov, Algebra i Logika **7**, No. 3 (1968), 63–104] for establishing faithful representability of groups by matrices. An exposition of this method can be found in Part 3 of [30].

17.2.6. Exercise. Let p be any prime number. Every finitely generated nilpotent torsion-free group is residually a finite p-group. (Hint. See the proof of Theorem 14.2.2.)

We promised at the beginning of this subsection that we would show that any finitely generated, nilpotent group G can be embedded in $\mathbf{SL}_n(\mathbf{Z})$ for some n (depending on the group); using Theorems 17.2.2 and 17.2.5 we shall now fulfil our promise. It suffices to find an embedding $\phi: G \to \mathbf{GL}_m(\mathbf{Z})$, since then the map defined by $x \to \begin{pmatrix} x^\phi & 0 \\ 0 & x^\phi \end{pmatrix}$ will serve as the desired embedding (into $\mathbf{SL}_{2m}(\mathbf{Z})$. To get ϕ we make use of the following general observation. Let H be a subgroup of finite index m in an arbitrary group G, and let $\{a_1, \ldots, a_m\}$ be a complete set of right coset representatives for H in G. If σ is a faithful representation of H by matrices of degree n, then the map defined by

$$g \to ((a_i g a_j^{-1})^\sigma),$$

where we set $x^\sigma = 0$ for $x \notin H$, is a faithful representation of G by matrices of degree mn. Clearly this is nothing more than the representation of G induced by the representation σ (see §13.2).

17.2.7. Exercise. If a group is almost representable by matrices over a ring with a multiplicative identity, then it is representable by matrices over that ring.

17.2.8. Exercise. A finitely generated nilpotent group is residually finite.

17.2.9. Exercise. There exist finitely generated, torsion-free nilpotent groups A, B each of which embeds in the other, but which are not isomorphic: for example take

$$A = \begin{pmatrix} 1 & \mathbf{Z} & \mathbf{Z} \\ 0 & 1 & \mathbf{Z} \\ 0 & 0 & 1 \end{pmatrix}, \quad B = \begin{pmatrix} 1 & n\mathbf{Z} & \mathbf{Z} \\ 0 & 1 & \mathbf{Z} \\ 0 & 0 & 1 \end{pmatrix} \quad \text{where } n \neq 0, 1.$$

17.2.10. Exercise. A finitely generated nilpotent group with finite center is itself finite.

17.3. Torsion-free Nilpotent Groups

The richest and most interesting part of the theory of these groups is that concerned with root extraction, which we shall now examine. The starting point is Theorem 16.2.8, according to which extraction of nth roots in a torsion-free nilpotent group G is a "partial operation" on the group; i.e. it is single-valued but not everywhere defined. It turns out that such a group G can always be embedded in a divisible nilpotent group (i.e. one in which all roots of all elements exist), and further that any two nilpotent divisible closures of G (i.e. "minimal" divisible supergroups of G) are isomorphic and consist entirely of roots of elements of G.

We recall the precise definition of divisibility: A group G is *divisible* if for every element g and every positive integer m the equation $x^m = g$ can be solved in G. As examples of divisible nilpotent groups we may once again take the groups $\mathbf{UT}_n(K)$ where K is any field of characteristic zero. To see this we extend formula (3) above by defining, for each $a \in \mathbf{UT}_n(K)$, $\mu \in K$,

$$a^\mu = \sum_{i=0}^{n-1} \binom{\mu}{i}(a-e)^i, \text{ where}\binom{\mu}{i} = \frac{\mu(\mu-1)\cdots(\mu-i+1)}{i!}, \qquad \binom{\mu}{0} = 1.$$

It is easy to verify that $a^{\mu+\nu} = a^\mu a^\nu$, $a^{\mu\nu} = (a^\mu)^\nu$. Hence in particular $a^{1/m}$ is an mth root of a; i.e. the group $\mathbf{UT}_n(K)$ is divisible.

Let G be a torsion-free, nilpotent group. We call a divisible, torsion-free nilpotent group a *(nilpotent) divisible closure* of G if it contains G but has no proper divisible subgroups containing G.

17.3.1. Theorem. *Let H be a subgroup of a divisible, torsion-free nilpotent group G. The set $\sqrt{H}$ of all elements of G some powers of which lie in H (the "radical closure" of H in G), is a subgroup of G, and therefore a divisible closure of H. The higher centers of the subgroups H and $\sqrt{H}$ are related in the following way:*

$$\zeta_i\sqrt{H} = \sqrt{(\zeta_i H)}, \qquad \zeta_i H = H \cap \sqrt{(\zeta_i H)}.$$

PROOF. (i) We shall first prove that $\sqrt{H}$ is a subgroup; for this it suffices to show that $xy \in \sqrt{H}$ for all x, $y \in \sqrt{H}$, since $\sqrt{H}$ is clearly closed under taking inverses. Set $A = \langle x, y \rangle$, $B = A \cap H$ and $A_i = \gamma_i A$. The desired conclusion $(xy \in \sqrt{H})$ will follow if we can show that $|A:B| < \infty$; to prove that $|A:B| < \infty$ we shall show that $|BA_i : BA_{i+1}| < \infty$ for all i. The series

$$A = BA_1 \geq BA_2 \geq \cdots$$

is subnormal with abelian factors, since by the commutator identities (3) of §3, Ch. 1,

$$[BA_i, BA_i] \leq [B, B]^{A_i}[A_i, B]^{A_i}[A_i, A_i] \leq BA_{i+1}.$$

Let m be a positive integer such that x^m, $y^m \in B$, and suppose as inductive hypothesis that BA_{i-1}/BA_i is finite of exponent m^{i-1} (this being easy for $i = 2$). We shall prove that then BA_i/BA_{i+1} is finitely generated and of exponent m^i. That BA_i/BA_{i+1} is finitely generated follows from the fact that A is polycyclic (Theorem 17.2.2). For the statement about the exponent, observe first that since

$$[A_{i-1}, A, A] \equiv 1 \pmod{A_{i+1}},$$

the function $f(u, v) = [u, v] \pmod{A_{i+1}}$, $u \in A_{i-1}$, $v \in A$, is homomorphic in both arguments (once again by the commutator identities!). Hence

$$A_i^{m^i} = [A_{i-1}, A]^{m^i} \leq [A_{i-1}^{m^{i-1}}, A^m]A_{i+1}.$$

Since

$$A_{i-1}^{m^{i-1}} \leq (B \cap A_{i-1})A_i, \qquad A^m \leq BA_2,$$

a last application of the commutator identities gives, finally,

$$A_i^{m^i} \leq BA_{i+1},$$

from which it follows that BA_i/BA_{i+1} has exponent m^i.

(ii) To establish the equations involving the higher centers we use induction. Proceeding immediately to the inductive step, assume their truth for i. Write briefly $H_i = \zeta_i H$. The inclusion $\zeta_{i+1}\sqrt{H} \leq \sqrt{H_{i+1}}$ follows from the fact that for every $x \in \zeta_{i+1}\sqrt{H}$ we have $x^m \in H$ for some m, so that using the inductive hypothesis,

$$[x^m, H] \leq H \cap \zeta_i\sqrt{H} = H \cap \sqrt{H_i} = H_i,$$

whence $x^m \in H_{i+1}$, $x \in \sqrt{H_{i+1}}$.

The reverse inclusion, $\sqrt{H_{i+1}} \leq \zeta_{i+1}\sqrt{H}$, is equivalent to elementwise commutativity modulo $\zeta_i\sqrt{H} = \sqrt{H_i}$, of $\sqrt{H}$ and $\sqrt{H_{i+1}}$. To show this take $x \in \sqrt{H}$, $y \in \sqrt{H_{i+1}}$; then $x^m \in H$, $y^n \in H_{i+1}$ for some positive integers m, n. Since

$$[H, H_{i+1}] \leq H_i \leq \sqrt{H_i},$$

we have that $x^m y^n \equiv y^n x^m \pmod{\sqrt{H_i}}$. Since $\sqrt{H}/\sqrt{H_i}$ is torsion-free (by Exercise 16.2.10) it follows (by Exercise 16.2.9) that $xy \equiv yx \pmod{\sqrt{H_i}}$, as required.

Finally we show that $H \cap \sqrt{H_{i+1}} = H_{i+1}$, or rather that $H \cap \sqrt{H_{i+1}} \leq H_{i+1}$, since the reverse inclusion is trivial. Let $x \in H \cap \sqrt{H_{i+1}}$, $x^m \in H_{i+1}$. Thus x^m commutes modulo H_i with every element of H, and therefore (by Exercises 16.2.9 and 16.2.10) so does x. Hence $x \in H_{i+1}$, and the proof is complete.

17.3.2. Theorem (A. I. Mal'cev). *Every torsion-free nilpotent group G has a nilpotent divisible closure of the same nilpotency class. Any two nilpotent divisible closures of G are isomorphic; moreover, given any automorphism ϕ of G there is an isomorphism between the two divisible closures which extends ϕ.*

PROOF. (i) *Uniqueness.* Let G_1, G_2 be isomorphic copies of G and let $\phi: G_1 \to G_2$ be an isomorphism between them. Suppose we have nilpotent divisible closures, denoted by $\sqrt{G_1}$, $\sqrt{G_2}$, of G_1, G_2 respectively. Form their direct product $P = \sqrt{G_1} \times \sqrt{G_2}$ and consider the subgroup $D = \{(x, x^\phi) \mid x \in G_1\}$, or more particularly its radical closure $\sqrt{D}$ in P. We shall show first that

$$P = (\sqrt{D}) \cdot (\sqrt{G_i}), \qquad (\sqrt{D}) \cap (\sqrt{G_i}) = 1 \quad \text{for } i = 1, 2. \tag{9}$$

If $x \in (\sqrt{D}) \cap (\sqrt{G_i})$ then $x^m \in D \cap G_i$ for some positive integer m; but then $x = 1$, proving the second of the equalities (9). It follows that the restriction to $\sqrt{D}$ of the projection map $\pi: P \to \sqrt{G_1}$, is a monomorphism, whence $(\sqrt{D})^\pi = \sqrt{(D^\pi)}$. Since $(\sqrt{G_1})^\pi = \sqrt{G_1}$ and $D^\pi = G_1$, it follows that $(\sqrt{D})^\pi = \sqrt{G_1}$. Hence $(\sqrt{D}) \cdot (\sqrt{G_2}) = P$. By symmetry we get $(\sqrt{D}) \cdot (\sqrt{G_1}) = P$, concluding the proof of (9).

The equalities (9) show that each element x_1 of $\sqrt{G_1}$ is the first component of exactly one element x of $\sqrt{D}$, and similarly for $\sqrt{G_2}$. Hence the map sending each x_1 to the projection of x on $\sqrt{G_2}$ is an isomorphism between $\sqrt{G_1}$ and $\sqrt{G_2}$, extending ϕ.

(ii) *Existence.* If the group G is finitely generated then, by Theorem 17.2.5, it can be embedded in the divisible nilpotent group $\mathbf{UT}_n(\mathbf{Q})$ for some n. The radical closure $\sqrt{G}$ of the subgroup G in the group $\mathbf{UT}_n(\mathbf{Q})$ will then be a nilpotent divisible closure of G. For each $g \in G$ and each positive integer m, we denote by ${}^m\sqrt{g}$ the (unique) solution in $\sqrt{G}$ of the equation $x^m = g$. It is obvious that

$$ {}^m\sqrt{g} = {}^n\sqrt{g_1} \Leftrightarrow g^n = g_1^m. \tag{10}$$

Also, by Part (i) of the proof, the multiplication table of $\sqrt{G}$ is completely determined by that of G. We now drop our assumption that G is finitely generated, and consider the set of formal symbols ${}^m\sqrt{g}$, $g \in G$, $m = 1, 2, \ldots$. The double implication (10) gives us an equivalence relation on this set of symbols: we denote by $\sqrt{G}$ the set of equivalence classes of this relation. We make $\sqrt{G}$ a group (containing G) by multiplying ${}^m\sqrt{g}$ and ${}^n\sqrt{g_1}$ (or rather their classes) just as the corresponding elements of the divisible closure of $\langle g, g_1 \rangle$ are multiplied. If the group G has nilpotency class s then the divisible closures of its finitely generated subgroups will have class $\leq s$ (by Theorem 17.3.1). Hence $\sqrt{G}$ satisfies the law $[x_1, \ldots, x_{s+1}] = 1$, and is therefore nilpotent of the same class as G. This completes the proof of the theorem.

A nilpotent group having nontrivial elements of finite order need not be embeddable in a divisible nilpotent group, since the torsion subgroup of a divisible nilpotent group necessarily lies in its centre. This can be seen as follows. Let $g \in G$ have finite order $m > 0$. Omitting the routine preliminaries of the induction, we may suppose that g lies in the second center of G. Then $[g, x^m] = [g, x]^m = [g^m, x] = 1$ for all $x \in G$. Since G is divisible x^m ranges over the whole of G as x does. Hence g is actually central.

§18. Generalizations of Nilpotency

Of the many generalizations of nilpotency we shall consider only three: local nilpotence, the normalizer condition, and the Engel condition.

18.1. Local Nilpotence

A group is said to be *locally nilpotent* if all of its finitely generated subgroups are nilpotent. A group which is locally nilpotent need not itself be nilpotent: consider for example the direct product of a sequence of nilpotent groups with nilpotency classes increasing to infinity, or the group $UT_\omega(K)$ of Exercise 17.3.1. More generally if σ is a group property inherited by subgroups, we say that a group G has the property σ *locally* if its finitely generated subgroups all have the property σ. (This general notion will play a larger role in the next chapter.)

18.1.1. Exercise. Subgroups and factor groups of locally nilpotent groups are again locally nilpotent. A locally nilpotent group is locally polycyclic.

18.1.2. Theorem. (*i*) *In any group the product of two normal, locally polycyclic subgroups is again locally polycyclic.*

(*ii*) (B. I. Plotkin) *In any group the product of two normal, locally nilpotent subgroups is again locally nilpotent.*

PROOF. (i) Let K, L be normal, locally polycyclic subgroups of an arbitrary group G. We wish to show that every finitely generated subgroup of KL is polycyclic. Take any finitely generated subgroup of KL together with a finite set of generators for it, and express each of the generators in the form ab, $a \in K$, $b \in L$. Let A be the subgroup generated by the left factors, and B by the right. The groups A, B are polycyclic since finitely generated. It clearly suffices to show that $H = \langle A, B \rangle$ is polycyclic. For this we shall use Lemma 3.2.9, which describes the structure of H in terms of A, B (note that, in the terminology of that lemma, the sets I, J, C are finite). In view of that lemma and Exercise 4.4.3 (from which it easily follows that the product of two normal polycyclic subgroups of a group is again polycyclic), it suffices to prove that the group $A[A, B]$ is polycyclic.

Since K, L are normal in G, we have that $[K, L] \leq K \cap L$. Since A, $\langle C \rangle$ are finitely generated and are contained in the locally polycyclic group K, it follows that $\langle A, C \rangle$ is polycyclic; hence its subgroup $\langle C^A \rangle$ is also polycyclic. On the other hand this latter group is contained in L, whence $\langle C^A, B \rangle$ is a finitely generated subgroup of L, and so polycyclic. By the lemma referred to above $[A, B] \leq \langle C^A, B \rangle$, so that $[A, B]$ is also polycyclic and therefore finitely generated. Hence $A[A, B]$ is a finitely generated subgroup of K and therefore polycyclic.

(ii) is proved in the same way.

Having proved this theorem it is appropriate, just as it was after Fitting's theorem, to refer the reader to Exercise 16.2.13.

As might perhaps be expected, not all of the properties of nilpotent groups revealed in §16 are possessed by locally nilpotent groups: for instance Example 18.2.2 of the next subsection shows that it is not always the case that every subgroup of a locally nilpotent group is subnormal. How much of Theorem 16.2.2 (asserting the subnormality of every subgroup of a nilpotent group) can be salvaged under the weaker assumption of *local nilpotence*? There is at least the following theorem.

18.1.3. Theorem (D. H. McLain). *Every maximal subgroup of a locally nilpotent group is normal.*

PROOF. Let H be a maximal subgroup of a locally nilpotent group G, and suppose H is not normal in G. Then there exists $x \in [G, G]$, $x \notin H$. By the maximality of H, we have that $\langle x, H \rangle = G$. Express x as a product of commutators:

$$x = [y_1, z_1] \cdots [y_n, z_n], \qquad y_i, z_i \in G.$$

The elements y_i, z_i can be expressed as words in x and a certain finitely many elements $h_1, \ldots, h_m \in H$. Set

$$H^* = \langle h_1, \ldots, h_m \rangle, \qquad G^* = \langle h_1, \ldots, h_m, x \rangle.$$

By hypothesis G^* is nilpotent. Clearly $x \in [G^*, G^*]$, $x \notin H^*$, so that H^* is a proper subgroup of G^*. By Theorem 16.2.2, H^* is subnormal in G^*, whence by Theorem 17.2.2 there is a polycyclic series from H^* to G^*:

$$H^* < H_1 < \cdots < H_s < G^*. \tag{1}$$

Since the quotient G^*/H_s is abelian, we must have $[G^*, G^*] \le H_s$, whence $x \in H_s$; but then $G^* \le H_s$, contradicting the strictness of the inclusions in (1). This completes the proof.

18.2. The Normalizer Condition

It has been shown by H. Heineken and I. J. Mohamed [A group with trivial center satisfying the normalizer condition, J. Algebra **10** (1968), 368–376] that in general the converse of Theorem 16.2.2 is false; i.e. there exist non-nilpotent groups all of whose subgroups are subnormal. In the positive direction, note however the following result of J. E. Roseblade [J. Algebra **2** (1965), 402–412]: Corresponding to each positive integer n there are positive integers $r(n)$ and $s(n)$ such that if every $r(n)$-generator subgroup of a group is a member of a subnormal series of length $\le n$, then the group is nilpotent of class $\le s(n)$.

The *normalizer condition* is weaker than that of subnormality of all subgroups: it requires only that every proper subgroup differ from its normalizer.

18.2.1. Theorem (B. I. Plotkin). *A group satisfying the normalizer condition is locally nilpotent.*

PROOF. Let G be a group satisfying the normalizer condition. By Zorn's lemma every element of G is contained in some maximal locally nilpotent subgroup H. It suffices to show that H is normal in G, since then we shall have G covered by locally nilpotent *normal* subgroups, so that by Theorem 18.1.2 the group G itself will be locally nilpotent. Write $N = N_G(H)$. If $g \in G$ is such that $N^g = N$, then H^g, as well as H, will be normal in N. By Theorem 18.1.2 it follows that the subgroup $H^g H$ is locally nilpotent. Hence by the maximality of H we have $H = H^g H$, whence $H = H^g$ so that $g \in N$. Thus N coincides with its normalizer. Since G satisfies the normalizer condition this implies that $N = G$, as required.

That the converse is false is shown by the following example (which incidentally settles some other questions).

18.2.2. EXAMPLE (M. I. Kargapolov). For each ordinal number α we define a group G_α inductively by:

$$G_0 = \mathbf{C}_p, \; G_\alpha = \begin{cases} \mathbf{C}_p \text{ wr } G_{\alpha-1} \text{ if } \alpha \text{ is not a limit ordinal;} \\ \bigcup_{\beta < \alpha} G_\beta \qquad \text{if } \alpha \text{ is a limit ordinal.} \end{cases}$$

It is clear from this construction that each G_α is a p-group which is locally finite (i.e. all of its finitely generated subgroups are finite) and satisfies $|G_\alpha| \geq |\alpha|$. Certainly, therefore, the G_α are locally nilpotent.

Denote by γ the first uncountable ordinal. Clearly all the G_α with $\alpha < \gamma$ are countable, while G_γ is uncountable. We shall show that G_γ does not satisfy the normalizer condition. Suppose the contrary. Then there is in G_γ a chain of subgroups H_μ indexed by the ordinals μ less than or equal to a certain ordinal ν, satisfying the following conditions:

(i) $H_0 = 1$; $H_\mu \lhd H_{\mu+1}$; $H_\mu = \bigcup_{\lambda < \mu} H_\lambda$ for limit ordinals μ; $H_\nu = G_\gamma$;

(ii) all the factors $H_{\mu+1}/H_\mu$ are cyclic.

(Such a chain may be constructed by defining inductively $H_{\mu+1}$ to be $\langle x, H_\mu \rangle$ for some x belonging to the normalizer of H_μ in G_γ, but outside H_μ, and by using the third of the statements in (i) to define H_μ for limit ordinals μ.) We shall obtain a contradiction from the putative existence of such a chain of subgroups of G_γ.

Clearly all the H_n indexed by natural numbers n are countable by (ii), so that their union H_ω is also countable. Enumerate the elements of H_ω in any way: $h_1, h_2, \ldots$. It is clear that $h_m \in G_{\beta_m}$ for some $\beta_m < \gamma$, and then that the union of these G_{β_m} is some G_β with $\beta < \gamma$ (since G_β is countable, while G_γ is not). We shall get our contradiction by showing (by induction on λ) that G_β

contains all H_λ with $\omega < \lambda \leq \nu$. This is obvious for limit ordinals λ (given its truth for smaller ordinals), so we assume λ is not a limit ordinal, and as inductive hypothesis that $H_{\lambda-1} \leq G_\beta$. Suppose that $H_\lambda \nleq G_\beta$; let $h \in H_\lambda$, $h \notin G_\beta$. There exists an α such that $h \in G_\alpha$, $h \notin G_{\alpha-1}$. Clearly $\beta \leq \alpha - 1$, whence $H_{\lambda-1} \leq G_{\alpha-1}$. By the definition of the restricted wreath product,

$$h = bf, \qquad b \in G_{\alpha-1}, \qquad f \in \mathbf{C}_p^{(G_{\alpha-1})},$$

where $f \neq 1$ since $h \notin G_{\alpha-1}$. For every $x \in H_{\lambda-1}$ we have

$$[x^b, f] = x^{-b} x^{bf} \in G_{\alpha-1} \cap \mathbf{C}_p^{(G_{\alpha-1})} = 1,$$

so that f centralizes $H_{\lambda-1}^b$. Thus the element $f' = bfb^{-1} \neq 1$ has infinite centralizer in $G_{\alpha-1}$. Now this centralizer, acting on $G_{\alpha-1}$ by right multiplication, must leave supp f' invariant. Since supp f' is finite, it follows that the centralizer is isomorphic to a group of permutations of a finite set, and so cannot be infinite.

18.2.3. Exercise. The group G_γ has trivial center.

Groups satisfying the normalizer condition are so close to groups with a (possibly infinite) central series, that for a long time it was unknown if a group with the normalizer condition can have trivial center. This hoary question was answered (in the affirmative) by Heineken and Mohamed (see the paper cited at the beginning of this subsection).

However it is still unknown ([25], Question 2.80) whether or not every group with the normalizer condition has a nontrivial abelian normal subgroup.

18.3. The Engel Condition

This generalization of nilpotency originates from the commutator law

$$[x_0, x_1, \ldots, x_n] = 1, \tag{2}$$

which, as we know, defines the variety of nilpotent groups of class $\leq n$. An attempt to restrict the number of variables involved leads naturally to the law

$$[x, \underbrace{y, \ldots, y}_{n}] = 1. \tag{3}$$

(We remind the reader that the commutators in (2) and (3) are left-normed; i.e. for example the left-hand side of (3) is $[\cdots[[[x, y], y], y] \cdots]$.) A group satisfying the law (3) is called a *boundedly Engel group of class* $\leq n$; this is in honour of F. Engel, who, together with Sophus Lie, laid the foundations of the theory of Lie groups and Lie algebras (see §16.1). It is obvious that the variety of all boundedly Engel groups of class n contains the variety of all nilpotent groups of class n. This containment is proper as the following example shows.

18.3.1. EXAMPLE (K. Weston). Let M be the set of positive integers that are "square-free", i.e. are not divisible by the square of any positive integer. With each $m \in M$ we associate a cyclic group of order 2 with generator a_m, and we denote by A the direct product of all the $\langle a_m \rangle$. For each prime p we define an automorphism ϕ_p of A by specifying its action on the generators of A as follows:

$$a_m^{\phi_p} = \begin{cases} a_m a_{m/p} & \text{if } p \mid m, \\ a_m & \text{if } p \nmid m. \end{cases}$$

Clearly ϕ_p^2 is the identity map, and $\phi_p \phi_q = \phi_q \phi_p$, so that the group Φ generated by all ϕ_p is abelian of exponent 2. Denote by G the extension of A by the group Φ of automorphisms (see §6.1). A direct computation gives that, for primes q not dividing m,

$$a_m = [a_{mq}, \phi_q].$$

It follows that $[G, G] = A$ and $[G, A] = A$, so that the group G is not nilpotent. On the other hand it satisfies the law $[x, y, y, y] = 1$, as we shall now show. For $a, a' \in A$; $\phi, \phi' \in \Phi$, we get successively that: $[a, \phi a'] = [a, \phi]$ (expand both sides and use the fact that A is abelian and normal; $[a, \phi, \phi] = 1$ (use the identity $[x, y, z] = [y, x][z, x][x, yz]$ and the fact that $[a, \phi]^2 = 1$); and finally $[\phi a, \phi' a', \phi' a', \phi' a'] = [\phi a, \phi' a', \phi', \phi'] = 1$, using the previous two equalities. Note incidentally the additional easy fact that Weston's group has exponent 4.

A group is said to be an *Engel group* if for each pair x, y of its elements the relation (3) is satisfied for some n depending on (x, y). It is clear that the class of Engel groups contains that of all locally nilpotent groups. However there are Engel groups which are not locally nilpotent, as the next example shows.

18.3.2. EXAMPLE (E. S. Golod). *For each $d \geq 2$ there exists a non-nilpotent d-generator group all of whose $(d-1)$-generator subgroups are nilpotent.* Each of these groups is constructed from an algebra with the analogous property, i.e. from an (infinite-dimensional) non-nilpotent algebra A on d generators with all of its $(d-1)$-generator subalgebras nilpotent. The details of the construction are set out in §23 of the Appendix, where A is given in the form F'/I, F' being the subalgebra of elements with zero constant term of the free associative algebra F over any field k, on the free generators $x_1, \ldots, x_d$, and I a certain homogeneous ideal contained in F'. For our example we take $k = \mathbf{GF}(p)$, where p is prime. Consider the following subset of F/I:

$$1 + A = \{1 + a \mid a \in A\}.$$

Clearly, $(1+a)(1+b) = 1 + a + b + ab$, and if $a^n = 0$ then

$$(1+a)^{-1} = 1 - a + a^2 - \cdots + (-1)^{n-1}a^{n-1},$$

so that $1 + A$ is a group. Further, since for each $a \in A$ there exists an m such that $a^{p^m} = 0$, it follows that

$$(1+a)^{p^m} = \sum_{i=0}^{p^m} \binom{p^m}{i} a^i = 1,$$

since all the binomial coefficients save the first and last are divisible by p. Thus $1 + A$ is a p-group. Let G be the subgroup generated by the elements $1 + \hat{x}_1, \ldots, 1 + \hat{x}_d$, where $\hat{x}_i = x_i + I$. It is clear that the subgroup generated by any $(d-1)$ elements of $1 + A$, say by $1 + a_1, \ldots, 1 + a_{d-1}$, $a_i \in A$, lies in $1 + A^*$, where A^* is the subalgebra generated by $a_1, \ldots, a_{d-1}$. By construction A^* is nilpotent, whence the group $1 + A^*$ is nilpotent (Exercise 16.1.4). Thus all $(d-1)$-generator subgroups of G are nilpotent. However G itself is *not* nilpotent. For in the contrary case G would be finite by Exercise 16.1.6. But then the group ring $k[G]$ and its natural homomorphic image $F/I = k \oplus A$ would also be finite, contradicting the fact that the algebra A is infinite. ·

It is still unknown whether or not a group must be locally nilpotent if for each ordered pair (x, y) of its elements the relation (3) holds for some n depending only on y. It is also unknown whether it is always the case that the product of two normal Engel subgroups of a group is again an Engel subgroup. A discussion of these problems, and some relevant results, may be found in [33].

7 Soluble Groups

§19. General Properties and Examples

19.1. Definitions

As already mentioned at the beginning of the preceding chapter, soluble groups are just those groups obtained as the result of finitely many successive extensions by abelian groups. The following theorem gives several useful alternative characterizations.

19.1.1. Theorem. *For an arbitrary group G the following four statements are equivalent*:

    (*i*) *The group G possesses a subnormal series with abelian factors*;

    (*ii*) *The group G possesses a normal series with abelian factors*;

    (*iii*) *The derived series $G \geq G' \geq G'' \geq \cdots \geq G^{(n)} \geq \cdots$ of the group G terminates in the identity after a finite number of steps*;

    (*iv*) *The group G satisfies one of the laws*

$$\delta_n(x_1, \ldots, x_{2^n}) = 1, \qquad n = 0, 1, 2, \ldots,$$

where

$$\delta_0(x) = x, \qquad \delta_{n+1}(x_1, \ldots, x_{2^{n+1}}) = [\delta_n(x_1, \ldots, x_{2^n}), \delta_n(x_{2^n+1}, \ldots, x_{2^{n+1}})].$$

PROOF. The implications (iii) $\Rightarrow$ (ii) $\Rightarrow$ (i) are clear.

(i) $\Rightarrow$ (iii). The hypothesis is that G has a subnormal series

$$1 = H_0 < H_1 < \cdots < H_s = G,$$

with abelian factors. Since G/H_{s-1} is abelian, we have $G' \leq H_{s-1}$. Suppose inductively that we already have the inclusion $G^{(k)} \leq H_{s-k}$ for some k with $1 \leq k < s$. Then from the commutativity of H_{s-k}/H_{s-k-1} we deduce that $(G^{(k)})' = G^{(k+1)}$ is contained in H_{s-k-1}, completing the inductive step. Hence $G^{(s)} = 1$.

So far we have established the equivalence of (i), (ii) and (iii). We now show that (iii) and (iv) are equivalent.

(iii) $\Rightarrow$ (iv). Suppose that the sth derived group $G^{(s)}$ is trivial. Since the $(s-1)$st derived group of $G/G^{(s-1)}$ is trivial we may suppose inductively that $\delta_{s-1}(g_1, \ldots, g_{2^{s-1}}) \in G^{(s-1)}$ for all $g_1, \ldots, g_{2^{s-1}} \in G$. The commutativity of $G^{(s-1)}$ then gives us that $\delta_s(g_1, \ldots, g_{2^s}) = 1$ for all $g_1, \ldots, g_{2^s} \in G$.

(iv) $\Rightarrow$ (iii). Suppose that the group G satisfies the law $\delta_s = 1$. From the definition of δ_s it follows that the verbal subgroup H of G determined by the word δ_{s-1} (i.e. the subgroup generated by the set of all values assumed by δ_{s-1} on G), is abelian. The quotient G/H satisfies the law $\delta_{s-1} = 1$, so by inductive considerations we may assume $G^{(s-1)} \leq H$. But then $G^{(s)} = 1$, as required. This completes the proof of the theorem.

We say that a group G is *soluble* if any of the statements (i), (ii), (iii), (iv) holds for it. A subnormal series with abelian factors is called a *soluble series*. If G is soluble then the least integer n such that $G^{(n)} = 1$ is termed the *solubility length* of G.

It is clear from the proof of Theorem 19.1.1 that the groups satisfying the law $\delta_n = 1$ are just those that are soluble of length $\leq n$. Consequently the class of soluble groups of length $\leq n$ is a variety. This observation and our knowledge of variety theory gives us immediately that subgroups and homomorphic images of soluble groups are soluble, and also that direct products of finitely many soluble groups or, more generally, Cartesian products of soluble groups of bounded length, are again soluble. Direct and Cartesian products of (infinitely many) soluble groups of unbounded soluble length will *not* be soluble.

19.1.2. Exercise. The group $\mathbf{T}_n(K)$ of (upper) triangular matrices is soluble. The direct product of the $\mathbf{T}_n(K)$, $n = 1, 2, \ldots$, is not soluble.

19.1.3. Exercise. An extension of a soluble group by a soluble group is again soluble.

19.1.4. Exercise. In any group the product of two soluble normal subgroups is again soluble.

19.1.5. Exercise. Every finite group G contains a (unique) soluble normal subgroup N, such that G/N has no nontrivial abelian normal subgroups.

19.1.6. Exercise. A finite soluble group has a subnormal series with factors cycle of prime order.

19.1.7. Exercise. A minimal normal subgroup of a finite soluble group is "elementary abelian," i.e. is the direct product of cyclic groups of the same prime order.

We remark without proof that whether or not a group is soluble is completely determined by the abstract structure of the lattice of its subgroups: if the subgroup lattice of a group G is isomorphic to that of some soluble group, then the group G is itself soluble. This result, the finite case of which was known previously (see for example Exercise 12, p. 178 of G. D. Birkhoff's "Lattice Theory," A.M.S. Colloquium Publications Vol. XXV, 1967), is due to B. V. Jakovlev [Algebra i Logika **9**, No. 3 (1970), 349–369].

19.2. Soluble Groups Satisfying the Maximal Condition

Polycyclic groups are obviously soluble. The subgroups of a polycyclic group, being themselves polycyclic (Exercise 4.4.3), are all finitely generated.

19.2.1. Exercise. A group G has all its subgroups finitely generated if and only if it satisfies the *maximal condition* (*for subgroups*): every ascending chain

$$H_1 \leq H_2 \leq \cdots$$

of subgroups is eventually constant; i.e. $H_n = H_{n+1} = \cdots$ for some n.

19.2.2. Exercise. The class of groups satisfying the maximal condition is closed under taking subgroups, homomorphic images, and extensions.

Soluble groups satisfying the maximal condition are easily described:

19.2.3. Theorem. *A soluble group satisfies the maximal condition if and only if it is polycyclic.*

PROOF. The sufficiency has already been noted. To prove necessity, suppose G is a group satisfying the maximal condition, with a soluble series

$$1 = G_0 \leq G_1 \leq \cdots \leq G_n = G.$$

Since the factors G_{i+1}/G_i also satisfy the maximal condition (see Exercise 19.2.2 above), they are finitely generated and therefore decompose as direct products of finitely many cyclic groups (Theorem 8.1.2). Hence the above series can be refined to a polycyclic series, so that G is polycyclic, as required.

A second obvious class of groups satisfying the maximal condition is that of finite groups. R. Baer [Noetherische Gruppen, I, Math. Zeitschr. **66**

(1956), 269–288] stated the following celebrated "maximal problem": Is every group satisfying the maximal condition a finite extension of a polycyclic group (in other words, an almost polycyclic group)? The answer is still unknown.†

An important subclass of the class of polycyclic groups is that of supersoluble groups: a group is *supersoluble* if it has a normal (not just subnormal) series with cyclic factors. As examples of supersoluble groups familiar to us the finitely generated nilpotent groups will serve.

19.2.4. Exercise. The subgroups and factor groups of a supersoluble group are supersoluble.

19.2.5. Theorem. *The commutator subgroup of a supersoluble group is nilpotent.*

PROOF. Suppose G is a supersoluble group and let

$$1 = G_0 \le G_1 \le \cdots \le G_n = G$$

be a normal polycyclic series for G. Let Z_{i+1}/G_i be the centralizer of the factor G_{i+1}/G_i in G/G_i. The intersection $\bigcap Z_i = Z$ say, centralizes all the factors of the above series, or in other words acts trivially on them by conjugation, so by Lemma 16.3.1, Z is nilpotent. By Remak's theorem (4.3.9) G/Z embeds in the direct product of the groups G/Z_i. Since G/Z_{i+1} is isomorphic to the group of automorphisms of G_{i+1}/G_i induced by the inner automorphisms of G/G_i, and since the automorphism group of a cyclic group is abelian, it follows that G/Z is abelian. Hence $[G, G] \le Z$, giving the theorem.

We end this subsection by establishing a property of *finite* supersoluble groups. Let G be a finite group of order $n = p_1^{\alpha_1} \cdots p_k^{\alpha_k}$, where the p_i are prime, and $p_1 > \cdots > p_k$. A normal series

$$1 = H_0 < H_1 < \cdots < H_k = G$$

is called a *Sylow series* of the group G, if for $i = 0, 1, \ldots, k-1$, the factor H_{i+1}/H_i is a Sylow p_{i+1}-subgroup of the group G/H_i. (Sometimes the requirement $p_1 > \cdots > p_k$ is omitted from the definition.)

19.2.6. Theorem. *A finite supersoluble group has a Sylow series.*

PROOF. Let p be the largest prime divisor of the order of the finite supersoluble group G, and let H be any normal subgroup of prime order, say q. The factor group G/H is supersoluble and $|G/H| < |G|$, so that using induction we may suppose that the Sylow p-subgroup P/H say, of G/H, is normal in G/H. If $p = q$ the subgroup P will be a Sylow p-subgroup of G and

---

† Now solved negatively: see Translator's Remarks, p. ix.

will also be normal in G. If $p \neq q$, then let H_1/H be a normal subgroup of G/H of order p. Since $p > q$ and $|H_1| = pq$, it follows from Sylow's theorem that there is just one Sylow p-subgroup H_2 of H_1, and then since $H_1 \trianglelefteq G$ we get that H_2 is also normal in G. By the inductive hypothesis, as applied to G/H_2, we then obtain the normality of the Sylow p-subgroup of G, which essentially completes the proof.

19.3. Soluble Groups Satisfying the Minimal Condition

We shall say that a group satisfies the *minimal condition* (*for subgroups*), if every descending chain $H_1 \geq H_2 \geq \cdots$ of subgroups eventually becomes stationary; i.e. $H_n = H_{n+1} = \cdots$ for some n. A group satisfying the minimal condition must be periodic, since the infinite cyclic group does not satisfy the condition.

19.3.1. Exercise. The class of groups satisfying the minimal condition is closed under taking subgroups, homomorphic images, and extensions.

In this subsection we shall elucidate the structure of soluble groups satisfying the minimal condition.

19.3.2. Theorem (S. N. Černikov). *Every soluble group satisfying the minimal condition is a finite extension of a direct product of finitely many quasicyclic groups.*

(This includes the case that the soluble group is finite, since then it is a finite extension of the direct product of the empty collection of groups, which is, of course, the trivial group.)

PROOF. Let G be an infinite soluble group satisfying the minimal condition. The minimal condition implies that G contains a subgroup H of finite index which itself has no proper subgroups of finite index. Since the intersection of two subgroups of finite index again has finite index it follows that there is exactly one such subgroup H, so that H is characteristic and therefore certainly normal. If H is abelian then it is divisible; to see this note that, for any prime p, $|H : H^p| < \infty$, so that $H^p = H$. By Theorem 9.1.6, H is therefore a direct product of quasicyclic groups. Since H satisfies the minimal condition, this direct product has only finitely many factors. This establishes the theorem in the case that H is abelian.

Suppose now that H is nonabelian. Let A be the last nontrivial term of the derived series of H. We shall show that A is contained in the center of H. Since A is abelian and satisfies the minimal condition, by the first part of the proof (which includes the case that G is abelian), A has only finitely many elements of each given order. Hence each $a \in A$ has only finitely many conjugates in H, so that $|H : C_H(a)| < \infty$. Since H has no proper subgroups of finite index, it follows that $C_H(a) = H$, whence $A \leq C(H)$, as required.

We deduce from this that H is nilpotent, by showing that if H is soluble of length k, then $H^{(k-i)} \leq \zeta_i(H)$. We have shown already that $A = H^{(k-1)} \leq \zeta_1(H)$. The same argument applied to the soluble group $H/H^{(k-1)}$ of length $k - 1$, yields $H^{(k-2)}/H^{(k-1)} \leq \zeta_1(H/H^{(k-1)})$, whence $H^{(k-2)} \leq \zeta_2(H)$, and so on.

Finally we prove that H must in fact be abelian. Let B be a maximal abelian normal subgroup of H. It follows as before that B is contained in the center of H. Yet by Theorem 16.2.6, B is its own centralizer in H. Hence $B = H$; i.e. H is abelian. This completes the proof.

We shall call any finite extension of a direct product of finitely many quasicyclic groups a *Černikov group*.

19.3.3. Exercise. The class of Černikov groups is closed under taking subgroups, homomorphic images, and extensions.

Since the automorphisms of the group

$$\mathbf{C}_{p^\infty} \times \cdots \times \mathbf{C}_{p^\infty} \ (n \text{ factors})$$

can be represented by matrices from $\mathbf{GL}_n(\mathbf{Z}_{p^\infty})$ (see Exercise 5.1.4), the next exercise turns out to be useful for the study of Černikov groups.

19.3.4. Exercise. The matrix group $\mathbf{GL}_n(\mathbf{Z}_{p^\infty})$ is almost torsion-free. It follows that, in particular, its periodic subgroups are finite. (Hint. The congruence subgroup modulo p^i, $i = 1, 2, \ldots$ (i.e. the group of matrices of the form $e + p^i a$, $a \in \mathbf{M}_n(\mathbf{Z}_{p^\infty})$), has finite index in $\mathbf{GL}_n(\mathbf{Z}_{p^\infty})$, and is torsion-free with the single exception $p = 2$, $i = 1$. The assertion about finite index is easy (cf. Exercise 4.2.7). The torsion-freeness follows without much difficulty from the binomial expansion

$$(e + p^i a)^m = e + \binom{m}{1} p^i a + \binom{m}{2} p^{2i} a^2 + \cdots + p^{mi} a^m.$$

S. N. Černikov posed the following "minimal problem": Is every group satisfying the minimal condition a Černikov group? (He used the term "extremal group.") In §22 we shall prove Černikov's theorem (19.3.2 above) under weaker assumption than solubility. We remark that the minimal problem has been shown to have a positive solution for locally finite groups [V. P. Šunkov, Algebra i Logika **9**, No. 2 (1970), 220–248]; however in its original generality the problem remains unsolved.†

† Now solved negatively: see Translator's Remarks, p. ix.

§20. Finite Soluble Groups

In the first subsection of this section we shall expound the theory of Hall and Carter subgroups of finite soluble groups, a theory which is strongly reminiscent of the Sylow theory of arbitrary finite groups. In the third and last subsection we shall give a criterion for supersolubility of a finite soluble group. The middle subsection is more auxiliary in nature, although the theorems of Maschke and Schur on arbitrary finite groups, which are included there, are very important in their own right.

20.1. Hall and Carter Subgroups

By Sylow's theorem (11.1.1), in an arbitrary finite group of order $n = p_1^{\alpha_1} \cdots p_s^{\alpha_s}$, where the p_i are distinct primes, there always exists, for each i, a subgroup of order $p_i^{\alpha_i}$, and any two such subgroups are conjugate. P. Hall discovered a generalization of this much of Sylow's theorem for finite *soluble* groups; to be precise he proved the existence and conjugacy of subgroups of order k of a finite soluble group of order n for those k such that $k \,|\, n$, $(k, n/k) = 1$. We shall call such divisors *Hall divisors* (of n). Any subgroup of a group whose order is a Hall divisor of the order of the group will be called a *Hall subgroup*. We shall prove Hall's theorem, but shall content ourselves with merely stating its converse: If a finite group G possesses Hall subgroups for all Hall divisors of $|G|$, then G is soluble. This is due to P. Hall [J. London Math. Soc. **12** (1937), 198–200], and S. A. Čunihin [Izv. NIIMM Tomskogo un-ta **2** (1938), 220–223].

20.1.1. Theorem (P. Hall). *Let G be a finite soluble group of order n and let k be a Hall divisor of n. Then*

    (*i*) *the group G contains at least one subgroup of order k;*

    (*ii*) *any two subgroups of G of order k are conjugate in G;*

    (*iii*) *any subgroup of order k' dividing k is contained in some subgroup of order k.*

PROOF. We use induction on n. The theorem being trivially true for the trivial group, suppose that $n > 1$ and, as inductive hypothesis, that the theorem is true for soluble groups of smaller order. Let A be a minimal normal subgroup of G. By Exercise 19.1.7, the subgroup A is the direct product of cyclic groups of prime order p; thus $|A| = p^m$ for some m.

If p divides k then the inductive hypothesis gives us the existence in the group G/A of a subgroup B/A say, of order k/p^m. Then B is a subgroup of the desired order k. Further, every subgroup of order k must contain A; hence if B_1 and B_2 are two subgroups of order k, we can form their quotients B_1/A and B_2/A. By the inductive hypothesis these groups are conjugate in G/A, so that B_1 and B_2 are conjugate in G.

Suppose now that p does not divide k. Denote by D the largest normal subgroup of G having order relatively prime to k, and by H/D a minimal normal subgroup of the group G/D. As before, we have that the order of H/D is q^s for some prime q, where $q^s | k$.

If the normalizer $N(Q)$ of a Sylow q-subgroup Q of H is the whole of G then of course $Q \triangleleft G$, and the statements (i) and (ii) follows as in the case $p | k$ (with Q, q in the roles of A, p). Thus we suppose $N(Q) \neq G$. From the fact that $G = N(Q)H = N(Q)D$ (see the Frattini lemma (17.1.8)) it follows that k divides the order of $N(Q)$. Hence by the inductive hypothesis, $N(Q)$, and therefore also G, contains a subgroup of order k. Let B_1, B_2 be two subgroups of G or order k. The intersections $B_1 \cap H$ and $B_2 \cap H$ are Sylow q-subgroups of H, and are therefore conjugate to Q; hence there exist conjugates $\hat{B}_1$, $\hat{B}_2$ of B_1 and B_2 such that $\hat{B}_1 \cap H = \hat{B}_2 \cap H = Q$. From the normality of H it then follows that Q is normal in $\langle \hat{B}_1, \hat{B}_2 \rangle$. By the inductive hypothesis $\hat{B}_1/Q$ and $\hat{B}_2/Q$ are conjugate in $\langle \hat{B}_1, \hat{B}_2 \rangle/Q$. Hence $\hat{B}_1$ and $\hat{B}_2$ are conjugate, and so therefore are B_1 and B_2.

Having established (i) and (ii), we now turn to the statement (iii). Let $\hat{B}$ be a subgroup of order k'. Let A be, as before, a minimal normal subgroup of order p^m. If $p | k$ then the order of the subgroup $A\hat{B}$ also divides k. By inductive hypothesis the group $A\hat{B}/A$ is contained in some subgroup of G/A of order k/p^m. It follows that $A\hat{B}$, and therefore $\hat{B}$, is contained in some subgroup of G of order k.

Suppose now that p does not divide k. As always the inductive hypothesis gives us that the subgroup $A\hat{B}/A$ of G/A is contained in some subgroup C/A of order k. We shall now work within the subgroup C which has order $p^m k$. By (i) C contains a subgroup, B say, of order k. Clearly the product AB has order $p^m k$ and so coincides with C. Hence certainly $A\hat{B}B = C$. From the equalities

$$|C| = \frac{|A\hat{B}| \cdot |B|}{|A\hat{B} \cap B|}, \qquad |C| = p^m k, \qquad |A\hat{B}| = p^m k', \qquad |B| = k,$$

we see that $D = A\hat{B} \cap B$ has order k'. It follows from statement (ii) of the theorem that $\hat{B}$ and D are conjugate in $A\hat{B}$, say $\hat{B} = D^g$. Thus $\hat{B}$ is contained in the subgroup B^g of order k. This completes the proof.

20.1.2. Exercise. In a finite soluble group G every maximal subgroup has prime-power index. For each prime divisor p of the order G, there is at least one maximal subgroup of G whose index is a power of p.

20.1.3. Exercise. If A is a Hall subgroup of a finite soluble group G and H is a subgroup containing $N_G(A)$, then $N_G(H) = H$.

In terms of the concept of a "Sylow basis", P. Hall was able to generalize his theorem, providing thereby the impulse for further generalizations in various directions. A collection $\{G_{p_1}, G_{p_2}, \ldots\}$ of Sylow p_i-subgroups of a group G is called a (*full*) *Sylow basis* for G if

(i) $G = \langle G_{p_1}, G_{p_2}, \ldots \rangle$,

(ii) $G_{p_i} G_{p_j} = G_{p_j} G_{p_i}$ for all i, j.

Sometimes in the definition of a Sylow basis the requirement (ii) is replaced by the following one:

(ii)′ For all $i_1, \ldots, i_s$ the prime divisors of orders of elements of the subgroup $\langle G_{p_{i_1}}, \ldots, G_{p_{i_s}} \rangle$ are precisely $p_{i_1}, \ldots, p_{i_s}$.

Usually the two definitions turn out to be equivalent.

Two Sylow bases $\{G_{p_i}\}$, $\{\hat{G}_{p_i}\}$ of a group G are said to be *conjugate* if there exists an element $g \in G$ such that for all i, $\hat{G}_{p_i} = G_{p_i}^g$.

P. Hall's generalization [On the Sylow systems of a soluble group, Proc. London Math. Soc. **43** (1937), 316–323] consists in the following: Every finite soluble group G has a Sylow basis, and any two such bases are conjugate in G. Any Sylow basis of a subgroup can be extended to a Sylow basis of the whole group G.

The concept of a Sylow basis makes sense for arbitrary periodic groups, so that it is natural to try to extend the theorem on the existence of Sylow bases to periodic soluble groups. Such extensions have been obtained for various classes of infinite groups; however there are periodic soluble groups which do *not* have a Sylow basis [M. I. Kargapolov, Some questions from the theory of nilpotent and soluble groups, Dokl. Akad. Nauk S.S.S.R. **127**, No. 6 (1959), 1164–1166; On generalized soluble groups, Algebra i Logika **2**, No. 5 (1963), 19–28].

A subgroup H of a group G is called a *Carter subgroup* of G if H is nilpotent and self-normalizing (i.e. coincides with its normalizer) in G.

20.1.4. Theorem (Carter). *A finite soluble group G has at least one Carter subgroup. Any two Carter subgroups of G are conjugate.*

PROOF. Once again we use induction on $|G|$. Suppose $|G| > 1$ and the theorem true for soluble groups of smaller order. Let A be a minimal normal subgroup of G. Then $|A| = p^m$ for some prime p. By inductive hypothesis G/A possesses a Carter subgroup, say $\hat{K}/A$. Let Q be a Hall p'-subgroup of $\hat{K}$; i.e. p does not divide $|Q|$, and $|\hat{K}| = |Q| \cdot p^l$. We shall show that $K = N_{\hat{K}}(Q)$ is a Carter subgroup of G.

By the generalized Frattini lemma (Exercise 17.1.9), we have $\hat{K} = N_{\hat{K}}(Q) \cdot QA = KA$. The subgroup K is the direct product of the nilpotent subgroup Q and $K \cap P$, where P is some Sylow p-subgroup of $\hat{K}$; hence K is nilpotent. Suppose g is such that $K^g = K$. Since $A \trianglelefteq G$ and $\hat{K} = KA$, we deduce that $\hat{K}^g = \hat{K}$, so that $g \in \hat{K}$. Since the normalizer of a Hall subgroup is self-normalizing (see Exercise 20.1.3), it follows that $g \in K$, as we wished to show.

Now let K_1, K_2 be two Carter subgroups of G. We shall show that they are conjugate in G.

First we prove that K_iA/A, $i = 1, 2$, are Carter subgroups of the group G/A. It is obvious that the K_iA/A are nilpotent. To see that they are self-normalizing, note first that if $(K_iA)^g = K_iA$ with $g \notin K_iA$, then since $|K_iA| < |G|$, it follows from the inductive hypothesis applied to the Carter subgroups of K_iA, that there exists an element $a \in K_iA$ such that $K_i^g = K_i^a$, or $K_i^{ga^{-1}} = K_i$. But then since K_i is self-normalizing we deduce that $ga^{-1} \in K_i$, contradicting the assumption $g \notin K_iA$.

Let Q_i be a Hall p'-subgroup of K_i. Then Q_i is also a Hall p'-subgroup of K_iA, so that $N_{K_iA}(Q_i)$ is, as before, nilpotent, and furthermore contains K_i. Hence $K_i = N_{K_iA}(Q_i)$.

By inductive hypothesis K_1A/A and K_2A/A are conjugate; we may therefore assume, by replacing K_2 by one of its conjugates, that $K_1A = K_2A$. But then Q_2, being a Hall p'-subgroup of K_2A, is a Hall p'-subgroup also of K_1A. Hence Q_1 and Q_2 are conjugate in KA_1. Therefore their normalizers in KA_1 are also conjugate. But in the preceding paragraph it was shown that these normalizers are just K_1, K_2. Hence K_1 and K_2 are conjugate, and the theorem is proved.

20.2. On the Complete Reducibility of Representations

Here we digress briefly into the theory of linear representations. Our chief goal in this subsection is the classical theorem of Maschke on complete reducibility of certain representations of finite groups, which we shall need in the next section. At the same time other facts about complete reducibility will emerge.

Let V be a vector space over the field K, and let G be a group of linear transformations of V. If V contains a proper subspace invariant under G, then G is said to be *reducible*. In the contrary case G is said to be *irreducible*, or to *act irreducibly* on the space V. The group G is *completely reducible* if every G-invariant subspace $U \subseteq V$ is complemented by a G-invariant subspace; i.e. if there exists a G-invariant subspace W such that $V = U \oplus W$.

A *linear representation* of an (abstract) group G is just a homomorphism ϕ from G to the group of invertible linear transformations of an n-dimensional vector space V over a field K. If the group G^ϕ of linear transformations is irreducible then we say that ϕ is *irreducible* (and similarly for the terms "reducible," "completely reducible"). A *matrix representation* is a homomorphism from G to $\mathbf{GL}_n(K)$. Needless to say the linear representations on the one hand, and the matrix representations on the other, enjoy an intimate relationship. This relationship is as follows: The linear representation ϕ yields for each fixed basis Σ of our space V, a matrix representation ϕ_Σ defined by taking g^{ϕ_Σ} to be matrix relative to the basis

Σ of the linear transformation $g\phi$. For a different basis Σ', we get in general a different matrix representation $\phi_{\Sigma'}$, which is *equivalent* to ϕ_Σ; i.e. $g^{\phi_{\Sigma'}} = t^{-1} g^{\phi_\Sigma} t$ for all $g \in G$, where t is the matrix of the change of basis from Σ to Σ' (and matrices, like maps, act on the right of row-vectors).

20.2.1. EXAMPLE. Let G be a finite group and A a normal subgroup which is elementary abelian; i.e. the direct product of n cyclic subgroups of order p. We turn the set A into a vector space over $\mathbf{GF}(p)$, the field of p elements, by defining addition and scalar multiplication in terms of the group operation as follows: $a + b = ab$, $\alpha a = a^\alpha$, where $\alpha \in \{0, 1, \ldots, p-1\} = \mathbf{GF}(p)$. It is easy to check that this does make A a vector space. Furthermore for each $g \in G$ the map $\hat{g}$ defined by $x \to x^g$, $x \in A$, is a linear transformation of the vector space A, and the map defined by $g \to \hat{g}$, is a homomorphism from G to the group of linear transformations of A. The upshot is that we have at our disposal a linear representation of G over $\mathbf{GF}(p)$. Incidentally, it is clear that $\hat{G} \simeq G/C_G(A)$.

20.2.2. Theorem (Maschke). *Let ϕ be a representation of the finite group G by linear transformations of the vector space V over the field K. If the order of G is not divisible by the characteristic of K, then ϕ is completely reducible.*

PROOF. The representation ϕ gives us an action of G on V; we shall write as usual, therefore, vg for $v(g^\phi)$ $(v \in V, g \in G)$. Let U by a G-invariant subspace of V. Let W be any subspace of V complementing U, so that $V = U \oplus W$. Denote by π_W the projection map of V onto W, and define a map $t: V \to V$ by setting

$$vt = \frac{1}{m} \sum_{g \in G} (vg^{-1})\pi_W g, \qquad v \in V,$$

where m is the order of G. It is easy to see that t is a linear transformation of V. For each fixed $h \in G$, the product gh ranges over the whole of the group G as g ranges over G. It follows from this that

$$(vt)h = \frac{1}{m} \sum_g (vh \cdot h^{-1}g^{-1})\pi_W gh = (vh)t,$$

so that the subspace $Vt = W_0$ say, is G-invariant.

We shall show that $V = U \oplus W_0$. First we express an arbitrary element v of V as a sum: $v = (v - vt) + vt$. The second term vt lies in W_0. The first term $v - vt$ is equal to

$$v - \frac{1}{m} \sum_g (vg^{-1})\pi_W g = \frac{1}{m} \sum_g (vg^{-1} - (vg^{-1})\pi_W)g.$$

Since $vg^{-1} - (vg^{-1})\pi_W \in U$, and U is G-invariant, it follows that $v - vt$ lies in U. Thus we have so far that $V = U + W_0$. Finally, suppose $v \in U \cap W_0$. Then since $v \in U$, since U is G-invariant, and since for any element u of U, the projection $u\pi_W = 0$, we have

$$vt = 0. \tag{1}$$

On the other hand, since v belongs to W_0 it is the image of some element $v' \in V$ under t; i.e.

$$v't = v. \tag{2}$$

But then $vt = v't^2$. Earlier we showed that $v' - v't \in U$, which implies that $v't - v't^2 = 0$. Hence $vt = v't$, and then (1) and (2) give us that $v = 0$, proving the theorem.

20.2.3. EXAMPLE. Let G, A be as in Example 20.2.1 above. Obviously a subgroup $B \leq A$ will be normal in G if and only if as a subspace of the vector space A, B is invariant under the group $\hat{G}$ of linear transformations of A. Let us assume that the prime p does not divide the order of the group $G/C_G(A)$, and that B is now a normal subgroup of G contained in A. Since the order of $|\hat{G}|$ is not divisible by p, we have by Maschke's theorem that $\hat{G}$ is completely reducible; hence the subspace B of A has a $\hat{G}$-invariant complement C. The set C is then a normal subgroup of the group G, and $A = B \times C$.

20.2.4. Exercise (part-generalization of Maschke's theorem). Let G be a finite group of linear transformations of a vector space V over a field of characteristic $p > 0$, and let P be a Sylow p-subgroup of G. If a G-invariant subspace U has a P-invariant complement, then it has a G-invariant complement.

(Solution. We have $V = U \oplus W$ where W is a P-invariant subspace. Denote by π_W the projection map of V onto W and define a map t from V to itself by

$$vt = \frac{1}{m} \sum_{g \in S} (vg^{-1})\pi_W g, \qquad v \in V,$$

where $m = |G:P|$, and S is a set of right coset representatives for P in G. It is easy to see that π_W commutes with the action of each element of P, so that the definition of the linear transformation t is independent of the choice of S. The proof of the G-invariance of the subspace $W_0 = Vt$, and that $V = U \oplus W_0$, now proceeds almost exactly as in the proof of Maschke's theorem.)

20.2.5. Exercise. Let A be an abelian normal subgroup of a finite group G, of order coprime to its index. If B is a direct factor of A, which is also normal in G, then there exists a normal subgroup D of G such that $A = B \times D$.

Maschke's theorem is a convenient tool for proving the following important result.

20.2.6. Theorem (Schur). *Suppose a finite group G has a normal subgroup A of order coprime to its index. Then A has a complement in G; that is, there exists a subgroup B of G such that $G = AB$, $A \cap B = 1$.*

PROOF (i) Suppose first that A is abelian of prime exponent p. By Theorem 6.2.8, we can embed G in the wreath product $W = A$ wr B, where $B = G/A$, in such a way that (regarding G as a subgroup of W)

$$W = G \cdot A^{[B]}, \qquad G \cap A^{[B]} = A$$

(see Exercise 6.2.9). Since $|G:A|$ is not divisible by p, and $A \lhd W$, it follows from Maschke's theorem that there exists a normal subgroup C of W, such that $A^{[B]} = A \times C$. From the equalities

$$W/C = (A^{[B]}/C) \cdot (BC/C), \qquad A^{[B]} \cap BC = C,$$

and the obvious isomorphism $\phi : W/C \to G$, we get that A is complemented in G by the image under ϕ of BC/C.

(ii) We shall now deal with the general case, using induction on the order of the group. Thus suppose the theorem false for G but true for groups of smaller order (i.e. that G is a "minimal counter-example" as they say). Write $k = |A|$ and $l = |G:A|$. Note that we have merely to produce a subgroup of G of order l, since the coprimality of k and l automatically implies that such a subgroup complements A in G.

Let P be a Sylow p-subgroup of A, and $N(P)$ its normalizer in G. Since $G = A \cdot N(P)$ (by the Frattini lemma (17.1.8)), the subgroup $N(P)$ contains as a normal subgroup of itself $A \cap N(P)$ having order coprime to its index l in $N(P)$. Hence if $|N(P)| < |G|$ then we should have by the inductive hypothesis that $N(P)$, and therefore also G, contains a subgroup of order l, contradicting our assumption about G. We deduce that the Sylow subgroups of A are normal in G, whence, certainly, $A' < A$. Suppose $A' \neq 1$. Then G/A' satisfies the hypotheses of the theorem, and $|G/A'| < |G|$. By the inductive hypothesis, G/A' contains, therefore, a subgroup C/A' of order l. Applying the inductive hypothesis to the group C we get the existence in it, and therefore in G, of a subgroup of order l. We are finally forced back on $A' = 1$, i.e. on A being abelian. Let p be a prime divisor of k. If $A^p \neq 1$ then by repeating the preceding argument with A^p in the role of A', we again get a contradiction. Hence A is abelian of exponent p. However Part (i) of the proof excludes this, we so we have reached our final contradiction, and the end of the proof.

It behoves us to mention a fact which nicely supplements Schur's theorem: The complements of the normal subgroup A are all conjugate. In the case that A is soluble, this was proved by H. Zassenhaus; and in the case that G/A is soluble, by S. A. Čunihin. But by the Feit-Thompson theorem on the solubility of groups of odd order (see beginning of §12), one or the other of the groups A, G/A must be soluble, so that the complements in Schur's theorem are conjugate.

We end this subsection with some general remarks about irreducibility and complete reducibility.

20.2.7. Lemma. *A group G of linear transformations of a finite-dimensional vector space V over a field K is completely reducible if and only if V is a direct sum of irreducible G-invariant subspaces.*

PROOF. Suppose G is completely reducible. Let U_1 denote any non-null, irreducible (i.e. minimal) G-invariant subspace of V. (Why does such a U_1 exist?) Suppose inductively that we have non-null irreducible G-invariant subspaces $U_1, \ldots, U_s$, which generate their direct sum $U_1 \oplus \cdots \oplus U_s = U$ say. If $U \neq V$, then by the complete reducibility there is a G-invariant subspace W complementing U. Take U_{s+1} to be a non-null irreducible G-invariant subspace of W; then $U_1, \ldots, U_{s+1}$ clearly generate their direct sum. Since this procedure must ultimately halt (certainly before we reach integers s with $s > \dim V$), induction yields the desired conclusion.

Conversely, suppose $V = U_1 \oplus \cdots \oplus U_r$, where the U_i are irreducible G-invariant subspaces. Let U be any G-invariant subspace. Let I be any subset of $\{1, 2, \ldots, r\}$ maximal with respect to the property

$$U \cap \bigoplus_{i \in I} U_i = 0.$$

Write $W = \bigoplus_{i \in I} U_i$. We shall show that W is the desired complement. Clearly $\langle U, W \rangle = U \oplus W$. If $v \neq U \oplus W$, then there exists a j such that $U_j \cap (U \oplus W) = 0$. But then $U \cap (U_j \oplus W) = 0$, contradicting the maximality of I.

20.2.8. Exercise. A group of linear transformations of a finite-dimensional vector space V is completely reducible if and only if every irreducible invariant subspace is complemented (by an invariant subspace).

20.2.9. Lemma. *Every finite-dimensional, irreducible representation of an abelian group over an algebraically closed field is one-dimensional.*

PROOF. Let G be an abelian group acting irreducibly on a finite-dimensional vector space V over an algebraically closed field K. Let λ be a characteristic root of the linear transformation $g \in G$, and let v be a corresponding eigenvector: $vg = \lambda v$, $0 \neq v \in V$. Then

$$(vf)g = (vg)f = \lambda(vf) \quad \text{for all } f \in G,$$

so that the set U of all eigenvectors of g corresponding to the eigenvalue λ constitutes a G-invariant subspace of V. By the irreducibility of G we must have $U = V$, so that every nonzero vector in V is an eigenvector of every $g \in G$. This and the irreducibility of G give the one-dimensionality of V.

20.3. A Criterion for Supersolubility

We shall prove the following result.

20.3.1. Theorem (Huppert). *A finite group is supersoluble if and only if every maximal subgroup has prime index.*

PROOF. *Necessity.* We shall suppose that G is finite and supersoluble and prove by induction on $|G|$ that every maximal subgroup has prime index.

Since G is supersoluble it contains for some prime p a normal subgroup of order p. If now M is a maximal subgroup of G, then either $M \cap N = 1$ or $N \leq M$. In the first case it is clear that M has index p. In the second case since M/N is maximal in G/N, by the inductive hypothesis $|G/N : M/N|$ is prime; but then so is $|G : M|$.

Sufficiency. Suppose the maximal subgroups of the finite group G have prime indices, and suppose that groups with smaller order than G satisfying this condition, are supersoluble. We begin by showing that G is soluble. Denote by p the largest prime divisor of $|G|$, and by $\hat{M}$ a maximal subgroup containing a Sylow p-subgroup of G. Then $|G : \hat{M}| < p$, so that by Exercise 11.3.6, the intersection $\bigcap_{g \in G} \hat{M}^g$ is nontrivial; it is also of course normal in G. Hence G has a proper normal subgroup, and therefore a minimal normal subgroup, different from itself. Let M be any such minimal normal subgroup, and let Q be a Sylow q-subgroup of M, where q is the largest prime dividing $|M|$. Suppose the normalizer $N(Q)$ of Q in G is not the whole of G. Since $N(Q) \cdot M = G$ by the Frattini lemma (17.1.8), we will certainly have $HM = G$ for H a maximal subgroup of G containing $N(Q)$. Hence $|G : H| = |M : M \cap H|$, so that $M \cap H$ has prime index $\hat{q}$ say, in M. This index also divides $|M : N_M(Q)|$ which is coprime to q, whence $\hat{q} < q$. Invoking once again Exercise 11.3.6, we deduce that in M there is a proper normal subgroup M_0 containing all q-elements (i.e. elements of q-power order) of M. Since M is normal, any conjugate of M_0 will have the same property, so that the intersection $\bigcap_{g \in G} M_0^g$ will be a proper subgroup of M, which is normal in G. Since this contradicts the minimality of the normal subgroup M, our assumption that $N(Q) < G$ must be false. Thus $Q \trianglelefteq G$, whence $Q = M$, so that M is certainly soluble. Since by the inductive hypothesis G/M is supersoluble, we conclude that g is soluble.

We know that a minimal normal subgroup of a finite soluble group is elementary abelian (Exercise 19.1.7). Thus since G is soluble this must be true for M, i.e. M is a direct product of q-cycles. It remains to prove that $|M| = q$, or at least that G contains a nontrivial cyclic normal subgroup. The proof splits into two cases.

(i) Suppose that the largest prime p dividing the order of G exceeds q. Since G/M is supersoluble (by the inductive hypothesis), it follows from Theorem 19.2.6 that it contains a normal subgroup A/M of order p. Let P be a subgroup of A of order p. If $P \triangleleft G$ then we have our desired cyclic normal subgroup, and thence the supersolubility of G. On the other hand if

$N(P) \neq G$, and if H is a maximal subgroup of G containing $N(P)$, then $H \ngeq M$, since otherwise we would have $N(P) \cdot A = HPM \leq H$. Hence $G = HM$, so that since M is abelian, we have $H \cap M$ normal in G. But then since M is minimal normal this intersection must be trivial, whence, $|G:H|$ being prime, the order of M must be q, as required.

(ii) Suppose that q is the largest prime dividing the order of G. In our present situation, the supersolubility of G/M together with Theorem 19.2.6 give that the Sylow q-subgroup G_q say, of G, is normal in G. From the consequent normality of the centre C of G_q in G, and the fact that $C \cap M \neq 1$ (M being a normal subgroup of the q-group G_q), we deduce that $M \leq C$.

If now $M = G_q$, then in the factor group G/M there is as before a normal subgroup A/M of prime order p, $p \neq q$. The supersolubility of G then follows as in Case (i).

Suppose that $M \neq G_q$, and let B/M denote a normal subgroup of G/M of order q. Since $M \leq C$, the group B is abelian. We may suppose that B has exponent q, i.e. is elementary abelian, since otherwise B^q would be a cyclic normal subgroup (of order q), and we would be finished.

Suppose as a first case that $B \nleq C$. Certainly B lies in the second centre of the group G_q. Since G/M is supersoluble, there exists a series

$$B = B_0 < B_1 < \cdots < B_s = G_q$$

where $B_i \trianglelefteq G$, $|B_{i+1}: B_i| = q$. Since $B \nleq C$ there is a number k such that $B \leq C(B_k)$, $B \nleq C(B_{k+1})$. Let $b \in B$, $b_{k+1} \in B_{k+1}$, be such that $B = \langle M, b \rangle$, $B_{k+1} = \langle B_k, b_{k+1} \rangle$. Then the set of all elements of the form $[b, x]$, $x \in B_{k+1}$, will be a subgroup of order q, normal in G. This can be seen as follows: Let $x = c_1 b_{k+1}^l$, $y = c_2 b_{k+1}^m$, where $c_1, c_2 \in B_k$, be any two elements of B_{k+1}, and let $g \in G$; then

$$[b, x][b, y] = [b, b_{k+1}^l][b, b_{k+1}^m] = [b, b_{k+1}]^{l+m},$$

$$[b, b_{k+1}]^g = [b^g, b_{k+1}^g] = [ab^n, c_3 b_{k+1}^r] = [b, b_{k+1}]^{n+r}.$$

where a, c_3 come from M, B_k respectively.

We come finally to the possibility that B is contained in the center of G_q. If we regard B as a vector space over $\mathbf{GF}(q)$, then the group F of automorphisms of B induced by conjugation by elements of G, becomes a group of linear transformations of that vector space (see Example 20.2.1). Since $B \leq C$, the order of $F \simeq G/C(B)$ is relatively prime to q. Hence by Maschke's theorem the invariant subspace M has an invariant complement A, which in addition must have dimension 1 since $|B/M| = q$. In group-theoretical language this means that A is a cyclic normal subgroup of G. Thus G always has a nontrivial cyclic normal subgroup, which, as noted before, completes the proof.

20.3.2. Exercise. A finite group is supersoluble if and only if its quotient by the Frattini subgroup is supersoluble.

§21. Soluble Matrix Groups

At the beginning of the chapter we noted (in Exercise 19.1.2) that the triangular matrix groups are soluble. As a source of other examples of soluble matrix groups we may take the finite soluble groups; for by Cayley's theorem (13.1.1) a finite group can be represented by permutations of a finite set, and thence by "permutation" matrices. Uniting these two kinds of examples into one, we end up with the class of extensions of triangular matrix groups by finite soluble groups: by Exercise 17.2.7 such groups are also representable by matrices. It turns out that, at least if the field is algebraically closed, the converse is also true: any soluble group of matrices over an algebraically closed field is conjugate in the full linear group to an almost-triangular group of matrices. This result, due to Kolchin and Mal'cev, will be the goal of the first subsection, §21.1. In the second, §21.2, the soluble subgroups of $\mathbf{GL}_n(\mathbf{Z})$ will be investigated; we shall see that they are all polycyclic. Finally, in §21.3, we shall prove that the holomorph of an arbitrary polycyclic group can be embedded in $\mathbf{GL}_n(\mathbf{Z})$, for appropriate n.

21.1. Almost-Triangularizability

A matrix group $H \leq \mathbf{GL}_n(K)$ will be called *triangularizable* if for some $g \in \mathbf{GL}_n(K)$, the group H^g is triangular. Before launching into the proof of the almost-triangularizability of soluble matrix groups we shall prepare the way by proving some general facts about irreducible and completely reducible matrix groups (see §20.2 for the definitions of these concepts and their elementary properties). In what follows we shall for the most part identify a matrix group G with its action; i.e. with a group of linear transformations of some vector space, the elements of G being actually the matrices of these linear transformations relative to some basis.

21.1.1. Theorem (Clifford). *A normal subgroup H of a completely reducible group G of linear transformations of a finite-dimensional vector space V, is itself completely reducible.*

PROOF. Let U_1 be an irreducible H-invariant subspace of V. There exists a set $\{g_1, \ldots, g_s\}$ of elements of G minimal relative to the property that the smallest G-invariant subspace $\hat{U}_1$ containing U_1 (in other words the closure of U_1 under G) is generated by the subspaces $U_i = U_1 g_i$, $i = 1, \ldots, s$. It follows from the normality of H that the U_i are H-invariant, and thence by the irreducibility of U_1 that $\hat{U}_1 = U_1 \oplus \cdots \oplus U_s$. If $\hat{U}_1 \neq V$, there is a G-invariant complement B say, of $\hat{U}_1$, containing a nonzero irreducible H-invariant subspace U_{s+1}. Repeating the above procedure with U_{s+1} in place of U_1 we get $\hat{U}_{s+1} = U_{s+1} \oplus \cdots \oplus U_{s+t} \subseteq B$. Clearly

$$\hat{U}_1 + \hat{U}_{s+1} = U_1 \oplus \cdots \oplus U_{s+t}.$$

Continuing in this way we shall after a finite number of steps have decomposed V into a direct sum of irreducible H-invariant subspaces. The
complete reducibility then follows from Lemma 20.2.7 above.

21.1.2. Lemma. *The center of an irreducible group of matrices over an
algebraically closed field consists of just the scalar matrices in the group.*

PROOF. Suppose a group G acts irreducibly on an n-dimensional vector
space V over an algebraically closed field. By Clifford's theorem the center
C of G is completely reducible, so that by Lemma 20.2.9 there is a basis
$v_1, \ldots, v_n$ of V consisting of characteristic vectors of every element of C.
Suppose there is an element $c \in C$ such that $v_i c = \lambda_i v_i$ with $\lambda_1 \neq \lambda_2$ (if $n = 1$
the lemma is trivial). Denote by U the subspace spanned by those v_i for
which $\lambda_i = \lambda_1$. This subspace is clearly proper; we shall prove that it is also
G-invariant, contradicting the irreducibility of G, and giving us the desired
conclusion.

Thus let g be any element of G, and express $v_1 g$ in terms of the basis:

$$v_1 g = \alpha_1 v_1 + \cdots + \alpha_n v_n.$$

From the equations

$$v_1(gc) = \alpha_1 \lambda_1 v_1 + \cdots + \alpha_n \lambda_n v_n,$$

$$v_1(cg) = \lambda_1(\alpha_1 v_1 + \cdots + \alpha_n v_n),$$

and $gc = cg$, we get that $\alpha_j = 0$ whenever $\lambda_j \neq \lambda_1$, whence $v_1 g \in U$. Here we
could have replaced v_1 by any $v_i \in U$, to conclude that for such v_i, $v_i g \in U$.
Hence U is G-invariant, as claimed.

The next lemma will enable us to induct on the degrees of matrices.

21.1.3. Lemma. *Suppose that the irreducible group G of matrices of degree n
over an algebraically closed field, contains a non-central abelian normal
subgroup H. Then G contains a reducible normal subgroup of finite index
$\leq (n!)!$*

PROOF. By Clifford's theorem the subgroup H is completely reducible. Thus
by Lemmas 20.2.7 and 20.2.9 we may suppose that H consists only of
diagonal matrices. Since H is not contained in the center of G, there exists
(by the preceding lemma) a non-scalar matrix $b \in H$. For every $g \in G$ the
element b^g lies in H and has the same characteristic roots as b. Hence
$|b^G| \leq n!$. The map $G \to \mathbf{S}(b^G)$ defined by

$$g \to \begin{pmatrix} b^x \\ b^{xg} \end{pmatrix},$$

is a homomorphism. Its kernel A is clearly contained in $C_G(b)$, and
$|G:A| \leq (n!)!$ (since $(n!)! \geq \mathbf{S}(b^G)$). The center of A contains the nonscalar
matrix b, so by the preceding lemma A cannot be irreducible.

21.1.4. Lemma. *If a group has an abelian normal subgroup of index n, then it has an abelian characteristic subgroup of finite index* $\leq n^n$.

PROOF. Let G be a group with an abelian normal subgroup A of index n. It is easy to see that there exist automorphisms $\theta_1, \ldots, \theta_s$ of G, $s \leq n$, such that the subgroup

$$S = \langle A^{\theta_1}, \ldots, A^{\theta_s} \rangle$$

is the same as that generated by all A^{θ}, $\theta \in \text{Aut } G$. The subgroup S will then, of course, be characteristic; moreover its center C contains $\bigcap A^{\theta_i}$, and therefore $|G:C| \leq n^n$. The subgroup C satisfies the requirements, and the proof is complete.

We are now almost ready for the proof of the theorem of Kolchin and Mal'cev. We lack only Exercise 24.3.2 of the Appendix, which can be solved by a direct application of one of the "local" theorems of logic. We recommend to the reader that he take that exercise on trust for the time being, and postpone studying the local theorems of logic till later: they will receive specific treatment in the next section (§22) of this chapter. Alternatively, solve the exercise directly!

21.1.5. Theorem (Kolchin–Mal'cev). *A soluble matrix group of degree n over an algebraically closed field contains a triangularizable subgroup of finite index less than some number depending only on n.*

PROOF (i) Suppose first that our soluble matrix group G is irreducible and nilpotent. By Lemma 20.2.9 it will suffice to prove that G has an abelian normal subgroup of finite index less than some number $\tau(n)$ depending only on n. We shall prove this by induction on n.

If G is nonabelian then it contains a non-central abelian normal subgroup (for instance any maximal abelian normal subgroup will serve: see Theorem 16.2.6). Thus by Lemma 21.1.3 G contains a reducible normal subgroup H of finite index $\leq (n!)!$. By Clifford's theorem the subgroup H is completely reducible and therefore, being also reducible, is a subdirect product of finitely many irreducible nilpotent groups H_i having degrees $<n$. By the inductive hypothesis each H_i contains an abelian normal subgroup of index less than $\tau(n-1)$; but then H must contain an abelian normal subgroup A of index less than $\tau(n-1)^n$. Hence $|G:A| < (n!)!\tau(n-1)^n$, and the intersection of the conjugates of A in G is then abelian normal of index less than $\{(n!)!\tau(n-1)\}!$ (see Poincaré's theorem (13.2.2)), which serves to define $\tau(n)$.

(ii) We next drop the assumption of nilpotency; i.e. we suppose only that G is irreducible (and of course soluble). We again show by induction on n that G possesses an abelian normal subgroup of finite index less than some number $\rho(n)$ depending only on n.

If G contains a non-central abelian normal subgroup, then the existence of an abelian normal subgroup of finite index follows exactly as in Part (i) of the proof. We may suppose therefore that the abelian normal subgroups of G are all central. Denote by R the subgroup generated by all nilpotent normal subgroups of G. By Part (i) and Exercise 24.3.2 of the Appendix, R contains an abelian normal subgroup of finite index $\leq \tau(n)$. By Lemma 21.1.4, R therefore contains an abelian *characteristic* subgroup B of finite index $\leq \tau(n)^{\tau(n)} = \tau_1(n)$ say. Since B is normal in G and abelian, it is central in G, and therefore certainly in R. Hence R is nilpotent. We shall now bound $|G:R|$.

Let

$$B = B_0 < B_1 < \cdots < B_s = R \tag{1}$$

be a central series of R whose terms are all characteristic in R (e.g. take the upper central series of R with the possible extra term B_0). Let Z be the centralizer of this series in G; i.e. $Z = \bigcap_{i=1}^{s} Z_i$, where Z_{i+1}/B_i is the centralizer in G/B_i of the factor B_{i+1}/B_i. It is not difficult to see that $|G:Z| \leq \tau_2(n)$, where $\tau_2(n)$ depends only on n. For, Z_{i+1}/B_i is the centralizer of B_{i+1}/B_i which has order $\leq |R:B| \leq \tau_1(n)$. Hence by Exercise 3.1.4, Ch. 1, we have that $|G/B_i : Z_{i+1}/B_i|$, which is the same as $|G:Z_{i+1}|$, is at most $\tau_1(n)^{\tau_1(n)}$. Then from Remak's theorem (4.3.9), it follows that

$$|G:Z| = \left| G : \bigcap_i Z_i \right| \leq (\tau_1(n)^{\tau_1(n)})^s \leq \tau_1(n)^{(\tau_1(n))^2},$$

which we may take as defining $\tau_2(n)$.

Thus Case (ii) will be disposed of once we have shown that $R = Z$; for then we may take $\rho(n) = \tau_1(n)\tau_2(n)$, with B the desired subgroup. Clearly $R \leq Z$. If $R < Z$, then since Z/R is soluble (being a factor of the soluble group G), it contains a nontrivial, abelian, characteristic subgroup A/R say; but then A is a nilpotent (throw it on the end of the series (1)), normal subgroup of G, properly containing R, which is impossible.

(iii) Finally, we consider the general case. Let V be a vector space and $\{e_1, \ldots, e_n\}$ a basis determining the action of G on V. Let

$$V = V_1 > V_2 > \cdots > V_{s+1} = 0 \tag{2}$$

be an unrefinable series of G-invariant subspaces. Thus G acts irreducibly on each factor V_i/V_{i+1}; therefore by Case (ii) G contains normal subgroups G_i, $1 \leq i \leq s$, such that $|G:G_i| \leq \rho(n)$, and the quotients G_i/A_i are abelian, where here A_i is the kernel of the induced action of G on V_i/V_{i+1}. Write $G_0 = \bigcap_i G_i$. Clearly G_0 is abelian, so that by Lemma 20.2.9, the series (2) can be refined to a G_0-invariant series

$$V = \hat{V}_1 > \hat{V}_2 > \cdots > \hat{V}_{n+1} = 0$$

with one-dimensional factors. If we use a basis $\{f_1, \ldots, f_n\}$ where $f_i \in \hat{V}_i \backslash \hat{V}_{i+1}$, to represent the linear transformations in G_0 by matrices, then these matrices will be triangular. Consequently the subgroup G_0 in its original form (as a subgroup of the matrix group G) is conjugate (by the matrix of the change of basis $\{f_j\}$ to $\{e_i\}$) to a subgroup of triangular matrices. Since the index of G_0 in G is bounded by some number depending only on n (use Remak's theorem again), the theorem follows.

21.1.6. Exercise. The solubility length of an arbitrary soluble group of matrices of degree n is less than some number depending only on n. Deduce that if a matrix group is locally soluble (i.e. if all its finitely generated subgroups are soluble) then it itself is soluble.

21.1.7. Exercise. A soluble matrix group of degree n possesses a normal subgroup whose commutator subgroup is nilpotent, and whose index is less than some number depending only on n. This fact is useful in particular for proving that abstract groups of one kind or another are not representable faithfully by matrices (see, for instance, [D. M. Smirnov, On generalized soluble groups and their group rings, Matem. sb. **67**, No. 3 (1965), 366–383]).

21.2. The Polycyclicity of the Soluble Subgroups of $\mathbf{GL}_n(\mathbf{Z})$

Let $\hat{\mathbf{Q}}$ denote the algebraic closure of the field $\mathbf{Q}$ of rational numbers. The elements of $\hat{\mathbf{Q}}$ are called *algebraic numbers*; it follows that an algebraic number is just a (complex) number satisfying some polynomial in one variable over the rationals. An *algebraic integer* is an algebraic number satisfying some polynomial over the integers with leading coefficient 1 (i.e. a *monic* polynomial over $\mathbf{Z}$).

Let K be a field of algebraic numbers (i.e. a subfield of $\hat{\mathbf{Q}}$) of finite degree over $\mathbf{Q}$. (Such a field K is an "algebraic number field.") In §25 in the Appendix we give a proof of the fact that the set k consisting of all algebraic integers in K, is a ring, which moreover has the properties that both it, as an additive group, and its (multiplicative) group of units (i.e. multiplicative invertibles) are finitely generated. With the help of these results from algebraic number theory, we shall now prove

21.2.1. Theorem (Mal'cev). *A soluble group of integral matrices is polycyclic.*

PROOF. Let G be a soluble group of integer matrices of degree n. By the Kolchin–Mal'cev theorem (21.1.5), there is a matrix $g = (g_{ij})$, and a subgroup $A \leq G$ of finite index, such that $B = A^g$ lies in the group $\mathbf{T}_n(K)$ of (upper) triangular matrices over the finite extension

$$K = \mathbf{Q}(g_{11}, g_{12}, \ldots, g_{nn})$$

of $\mathbf{Q}$. It follows from Exercise 25.1.1 of the Appendix that there exists $m \in \mathbf{Z}$ such that mg_{ij} is an algebraic integer for all i, j; hence by replacing g by mg in the above and renaming, it is clear that we may assume that the g_{ij} are algebraic integers in K. It then follows that the entries in the matrices of B can each be expressed as the quotient of an algebraic integer by the determinant of g (use the fact that g^{-1} is the adjoint matrix of g divided by det g). Let k denote the ring of all the algebraic integers in K, and, as usual, let k^* be the multiplicative group of k. The entries on the main diagonal of each matrix from B are roots of the characteristic polynomial of the corresponding matrix of A; since this polynomial has coefficients from $\mathbf{Z}$ and leading coefficient ± 1, it follows that the diagonal entries of matrices from B are algebraic integers, and, since their product is ± 1 they in fact lie in k^*. Consider the homomorphism

$$B \to \underbrace{k^* \times \cdots \times k^*}_{n}$$

which sends each matrix in B to its main diagonal; denote by C its kernel. By Theorem 25.1.12 of the Appendix, the quotient group B/C is finitely generated. Obviously the kernel C consists of all the unitriangular matrices in B. If we denote by C_i the subgroup of all matrices in C having their i diagonals immediately above the main diagonal all zero (i.e. $C_i = C \cap \mathbf{UT}_n^{i+1}(K)$), then the series

$$C = C_0 \geq C_1 \geq \cdots \geq C_{n-1} = 1$$

is clearly a normal series for C. It is easy to see that each factor C_i/C_{i+1} is isomorphic to a subgroup of the direct sum of $n - i - 1$ copies of the additive group of the ring k, which group is, by Theorem 25.1.6 of the Appendix, finitely generated. Hence C is finitely generated, whence also B, and therefore its conjugate A. Since the index of A in G is finite it follows that G is finitely generated. Thus every soluble subgroup of $\mathbf{GL}_n(\mathbf{Z})$ is finitely generated, so that every soluble subgroup is polycyclic, as required.

With the aid of this theorem we can prove the following stronger result.

21.2.2. Theorem. *A soluble group of automorphisms of a finitely generated abelian group is polycyclic.*

PROOF. Let A be a finitely generated abelian group and let Φ be a soluble subgroup of Aut A. Denote by H the torsion subgroup of A, and consider the homomorphism $\tau \colon \Phi \to \operatorname{Aut}(A/H)$, sending each $\phi \in \Phi$ to the automorphism $\hat{\phi}$ of A/H induced by ϕ; i.e. $(Ha)^{\hat{\phi}} = H(a^\phi)$. By the preceding theorem the factor group Φ/Ψ, where $\Psi = \operatorname{Ker} \tau$, is polycyclic. Hence we shall certainly have our theorem if we can show that Ψ is finite. To this end, first note that if $A = \langle a_1, \ldots, a_n \rangle$, then for any $\psi \in \Psi$, $a_i^\psi = h_i a_i$, $h_i \in H$. Since H is finite there are therefore only finitely many choices of images for each a_i. Since there are only finitely many a_i, it follows that Ψ is finite (and therefore certainly polycyclic, being soluble). This completes the proof.

We know (and have used above) that a soluble group is polycyclic if and only if its subgroups are all finitely generated (Theorem 19.2.3). We are now able to weaken this criterion as follows.

21.2.3. Theorem. *A soluble group is polycyclic if and only if all of its abelian subgroups are finitely generated.*

PROOF. We have the necessity, so we need only prove sufficiency. Thus let G be a soluble group with all of its abelian subgroups finitely generated. We shall suppose G has soluble length greater than 1, and, inductively, that the desired conclusion is valid for groups of shorter length. Let H be the last nontrivial member of the derived series of G, and suppose that the factor group G/H contains an abelian subgroup A/H which is not finitely generated. Since the factor group A/B, where $B = C_A(H)$, is abelian and isomorphic to a subgroup of Aut H, it is, by Theorem 21.2.2, finitely generated. Hence B/H cannot be finitely generated. Notice that since B/H is abelian, and H is contained in the center of B, the group B is nilpotent.

Let $\hat{H}$ be a maximal abelian normal subgroup of B, containing H. Since B is nilpotent, the factor group $B/\hat{H}$ embeds in Aut $\hat{H}$ (Theorem 16.2.6, Ch. 6). By hypothesis $\hat{H}$ is finitely generated, whence by Theorem 21.2.2 so is $B/\hat{H}$, and therefore also B, and then B/H, yielding a contradiction. Hence G/H does *not* have non-finitely generated abelian subgroups, and so, by the inductive hypothesis, is polycyclic. Since H is finitely generated abelian, the desired polycyclicity of G follows.

21.3. The Embeddability in $\mathbf{GL}_n(\mathbf{Z})$ of the Holomorph of a Polycyclic Group

Let G be a group. It is easy to check that the equation

$$(\textstyle\sum \alpha_i g_i)(\phi g) = \sum \alpha_i g_i^\phi g, \qquad \alpha_i \in \mathbf{Z}; \qquad \phi \in \text{Aut } G; \qquad g_i, g \in G, \quad (3)$$

defines an action of the holomorph Hol G on the integral group ring $\mathbf{Z}[G]$.

21.3.1. Lemma. *Let $G \leq \mathbf{GL}_n(\mathbf{Z})$, let N be a unitriangular normal subgroup of G, and suppose that G/N is finitely generated. If Φ is a subgroup of Aut G leaving N invariant and acting trivially on G/N, then there exists a monomorphism $\Phi G \to \mathbf{GL}_m(\mathbf{Z})$, which is unitriangular on N.*

PROOF. Extend the given embedding $G \leq \mathbf{GL}_n(\mathbf{Z})$ to a ring homomorphism $\rho: \mathbf{Z}[G] \to \mathbf{M}_n(\mathbf{Z})$, and let K be the kernel of ρ. Let I be the ideal in the ring $\mathbf{Z}[G]$ generated by all elements of the form $1 - x$, where $x \in N$. From the fact that N consists of unitriangular matrices it follows that for any n elements $x_1, \ldots, x_n \in N$,

$$((1 - x_1)(1 - x_2) \cdots (1 - x_n))^\rho = 0,$$

whence $I^n \subseteq K$, which in turn implies that $(I + K)^n \subseteq K$. Note that since N is finitely generated (being a subgroup of the finitely generated nilpotent group $\mathbf{UT}_n(\mathbf{Z})$—see Example 16.1.2), and G/N is finitely generated by hypothesis, we have that G is finitely generated, and therefore that the ring $\mathbf{Z}[G]$ is finitely generated. Also, as additive group $\mathbf{Z}[G]/K$ is finitely generated, since it is isomorphic to a subgroup of the finitely generated additive group $\mathbf{M}_n(\mathbf{Z})$. These two observations allow us to apply Theorem 23.1.4 of the Appendix, to deduce that $\mathbf{Z}[G]/(I + K)^n$ is finitely generated as additive group. Write $T/(I + K)^n$ for its torsion subgroup. Since the additive group $\mathbf{Z}[G]/K$ is torsion-free, we have that $T \subseteq K$. For any $\phi \in \Phi$, $g \in G$, we have by hypothesis that $g^\phi = gx$ for some $x \in N$, whence $g^\phi - g = g(x - 1) \in I$. It is straightforward to check, using this, that $I + K$ is invariant under the action of ΦG on $\mathbf{Z}[G]$ defined in (3). It follows that that action induces an action of the group ΦG on the finitely generated additive group $\mathbf{Z}[G]/T$. This action is faithful since $T \subseteq K$ (consider the action of $\phi g \in \Phi G$ on the elements of $G \subseteq \mathbf{Z}[G]$, which remain distinct modulo K).

It remains to verify that N acts on $\mathbf{Z}[G]/T$ unitriangularly. We are looking for a series

$$T < T_1 < \cdots < T_m = \mathbf{Z}[G]$$

of N-invariant subgroups of the additive group $\mathbf{Z}[G]$, such that its factors are infinite cyclic, and the induced action of N on each T_{i+1}/T_i is trivial. Since $I^n \subseteq T$, there is an $s \geq 0$ such that $I^s \not\subseteq T$, $I^{s+1} \subseteq T$ (where we set $I^0 = \mathbf{Z}[G]$). Then every element of $I^s \backslash T$ is, modulo T, fixed by N, so that there is a free generator of the free abelian group $\mathbf{Z}[G]/T$ fixed by N. Define T_1/T to be the subgroup generated by that free generator. Then T_2 is defined by repeating this argument with T_1 in place of T, and so on, by induction, for the other T_i. This concludes the proof of the lemma.

21.3.2. Theorem (Ju. I. Merzljakov). *The holomorph* Hol G *of any poly-cyclic group* G *can be embedded in* $\mathbf{GL}_n(\mathbf{Z})$ *for some* n *(depending on* G*).*

PROOF. (i) We first show that G possesses a characteristic, torsion-free subgroup M of finite index, such that, in addition, the quotient M/N of M by its (unique) maximal nilpotent normal subgroup N (prove its existence and uniqueness!) is torsion-free abelian. To begin with, take a normal series

$$G = G_1 > G_2 > \cdots > G_{l+1} = 1,$$

with each factor either torsion-free or of prime exponent. For each $i = 1, \ldots, l$, the action of G by conjugation on the factor G_i/G_{i+1} defines a matrix representation $\phi_i \colon G \to \mathrm{Aut}(G_i/G_{i+1})$ (relative to some basis for G_i/G_{i+1}). By the Kolchin–Mal'cev theorem (21.1.5), for those i such that G_i/G_{i+1} is torsion-free, G^{ϕ_i} contains a subgroup S_i of finite index, which is triangularizable in $\mathbf{GL}_{n_i}(\mathbf{C})$, where n_i is the rank of G_i/G_{i+1}. Write $T_i = S_i^{\phi_i^{-1}}$. For those i for which G_i/G_{i+1} has prime exponent, define $T_i = \mathrm{Ker}\ \phi_i$.

Then $\bigcap_{i=1}^{l} T_i$ has finite index in G (since each T_i does). By Exercise 17.2.3 this intersection contains a torsion-free subgroup T of finite index. Since for each i the subgroup T acts on G_i/G_{i+1} "triangularizably" (in fact trivially on those G_i/G_{i+1} of prime exponent), it follows that T', the commutator subgroup of T, acts "*uni*triangularizably" on the G_i/G_{i+1}. More precisely, $(T')^{\phi_i}$ is a unitriangularizable group of matrices, and so by Example 16.1.2 is nilpotent. In other words

$$[\underbrace{T', \ldots, T'}_{k_i}]^{\phi_i} = 1, \quad \text{for some } k_i.$$

Since G acts on G_i/G_{i+1} by conjugation, this is the same as saying that

$$[\underbrace{T', \ldots, T'}_{k_i}, G_i] \le G_{i+1}.$$

It follows that T' itself is nilpotent. We know from Exercise 15.2.3 that in a finitely generated group every subgroup of finite index contains a verbal subgroup of finite index; let V be such a verbal subgroup of G contained in T, and let N be the unique maximal nilpotent normal subgroup of V. Since $V' \le T' \cap V \le N$, the factor group V/N is abelian. Let m be the exponent of its torsion subgroup, and set $M = V^m N$. It is then clear that M and N have the required properties.

(ii) Let Φ be the group consisting of all those automorphisms of M which induce the identity automorphism on M/N. We shall show that there is a (faithful) matrix representation of the group ΦM, which is unitriangular on N. With this in view, we first prove by induction on the rank r of the free abelian group M/N, that M can be so represented. If $r = 0$ then $M = N$ is nilpotent, and by Theorem 17.2.5 can be embedded in $\mathbf{UT}_k(\mathbf{Z})$ for some k. Proceeding to the inductive step, suppose $r \ge 1$, and let $M_0 \ge N$ be such that M_0/N is free abelian of rank $r - 1$. Let $a \in M$ be such that $\langle a \rangle M_0 = M$; then $\langle a \rangle \cap M_0 = 1$ and $M_0 \lhd M$. Assuming inductively that M_0 has a matrix representation of the required sort, we get one for M from Lemma 21.3.1 above. A second application of that lemma then immediately gives us the desired representation of ΦM.

(iii) As the third step we show that $\operatorname{Hol} M$ can be faithfully represented by matrices over $\mathbf{Z}$. In view of Exercise 17.2.7 and Part (ii) above, this will follow from the fact that $|\operatorname{Aut} M : \Phi| < \infty$, which we shall now prove. Firstly, since M/N is abelian, it follows from the second of the commutator identities (3) of §3.2, Ch. 1, that in $\operatorname{Hol} M$ (or rather $\operatorname{Hol}(M/N)$)

$$[M/N, \alpha]^m \le [(M/N)^m, \alpha] \le ([M^m, \alpha]N)/N \tag{4}$$

for all $\alpha \in \operatorname{Aut} M$, $m = 1, 2, \ldots$. Let $\beta \in \operatorname{Aut} M$. Since the subgroup $\langle \beta \rangle M$ of $\operatorname{Hol} M$ is polycyclic, it contains, by Part (i) of the proof, a normal subgroup $\hat{T}$ of finite index m say, whose commutator subgroup $\hat{T}'$ is nilpotent. Since N is the largest nilpotent normal subgroup of M, we have that

$$[M^m, \beta^m] \leq \hat{T}' \cap M \leq N.$$

From this and (4) we get that $[M/N, \beta^m]^m = 1$. Since M/N is torsion-free, this yields in turn $[M/N, \beta^m] = 1$, that is, $\beta^m \in \Phi$, so that the factor group $(\operatorname{Aut} M)/\Phi$ is periodic. If r is the rank of M/N, we have an obvious monomorphism $(\operatorname{Aut} M)/\Phi \to \mathbf{GL}_r(\mathbf{Z})$. Since $\mathbf{GL}_r(\mathbf{Z})$ is almost torsion-free (see Exercise 19.3.4 of this chapter), we deduce that $(\operatorname{Aut} M)/\Phi$ is finite.

(iv) We are now ready to prove the theorem, namely that $\operatorname{Hol} G$ has a faithful matrix representation over $\mathbf{Z}$. We shall use the fact that for any group X there is an embedding

$$\operatorname{Hol} X \to \operatorname{Aut}(X \operatorname{wr} \mathbf{Z}_2). \tag{5}$$

This embedding comes from the following (faithful) action of $\operatorname{Hol} X$ on $X \operatorname{wr} \mathbf{Z}_2$: for all $\phi \in \operatorname{Aut} X$; $x, y, g \in X$, define

$$\begin{pmatrix} x & 0 \\ 0 & y \end{pmatrix} \phi g = \begin{pmatrix} (x^\phi)^g & 0 \\ 0 & y^\phi \end{pmatrix}; \quad \begin{pmatrix} 0 & x \\ y & 0 \end{pmatrix} \phi g = \begin{pmatrix} 0 & g^{-1}(x^\phi) \\ (y^\phi)g & 0 \end{pmatrix}. \tag{6}$$

It is straightforward to verify that the set of matrices like those in the left-hand sides of the equations in (6), form under multiplication a group isomorphic to $X \operatorname{wr} \mathbf{Z}_2$, and that those equations do define a faithful action of $\operatorname{Hol} X$. If X happens to be polycyclic, then $X \operatorname{wr} \mathbf{Z}_2$ will also be polycyclic, so that in view of (5) we need only show that $\operatorname{Aut} G$ has a faithful matrix representation over $\mathbf{Z}$. To reduce the problem further, let Γ be the group of all automorphisms of G inducing the identity map on the (finite) quotient G/M. Then since $|\operatorname{Aut} G:\Gamma| < \infty$, we need only show that Γ can be represented (again by Exercise 17.2.7). Let S be a finite subset of G which generates G modulo M. It can be verified directly that the map

$$\Gamma \to \operatorname{Hol} M \times \cdots \times \operatorname{Hol} M,$$

defined by

$$\gamma \to \prod_{s \in S} \gamma|_M \cdot [\gamma, s],$$

is a monomorphism. Since by (iii) $\operatorname{Hol} M$ has a faithful matrix representation over $\mathbf{Z}$, it follows that Γ can be so represented also, as required.

21.3.3. Exercise. If M is a characteristic subgroup of finite index in some group G, and $\operatorname{Hol} M$ has a faithful matrix representation, then so does $\operatorname{Hol} G$.

21.3.4. Exercise. The holomorph of an arbitrary polycyclic group is almost torsion-free, and for each prime p is almost a residually finite p-group. (Hint. Look at the proof of Theorem 14.2.2.)

The importance of Theorem 21.3.2 lies in the fact that it provides an avenue through which the methods of algebraic geometry, number theory, and p-adic analysis—depending on the ground ring of coefficients in which we choose to embed $\mathbf{Z}$—may be brought to bear on the study of polycyclic groups. For instance, as Wehrfritz has noted, the theorem can be used to prove that the automorphism group of an arbitrary polycyclic group G is finitely presented [L. Auslander, Ann. Math. **89** (1969), 314–322] in the following way. By Theorem 21.3.2 we may assume Hol $G \leq \mathbf{GL}_n(\mathbf{Z})$. If the group G is closed in the Zariski topology, then its normalizer and centralizer in $\mathbf{GL}_n(\mathbf{Z})$ are also closed, and therefore finitely presented [A. Borel, Proc. Internat. Congress of Mathns., Stockholm, 1962, 10–22]. Therefore the group Aut $G \simeq N_*(G)/C_*(G)$ is also finitely presented. If G is not closed the argument is somewhat more complicated. The details of this, and other interesting comments concerning Theorem 21.3.2, may be found in Wehrfritz' lectures [42], given on June 22, 1973, at the British Symposium held in London.

In connexion with the construction which we met in passing at the end of the proof of Theorem 21.3.2, it is appropriate to mention the following

21.3.5. Question ([25]). Let C be a fixed, nontrivial group (for example $C = \mathbf{Z}_2$). It is known [Algebra i Logika **9**, No. 5 (1970), 539–558], that for any groups A, B; all splitting extensions of B by A can be embedded in a certain uniquely defined way in the direct product $A \times \text{Aut}(B \text{ wr } C)$. How are they situated in there?

§22. Generalizations of Solubility

22.1. Kuroš–Černikov Classes

As we know, to say that a group is soluble can be taken as meaning that the group has a subnormal series with abelian factors, or alternatively a normal series with abelian factors, or again that some term of its derived series is trivial: all these conditions are equivalent. It is natural to define analogues of these (and other) conditions for "infinite subnormal series," and then to attempt to clarify the properties of, and relations between, the classes of groups possessing various such infinite series. (It seems realistic to expect that equivalent conditions defined in terms of ordinary (finite) series, will cease to be equivalent once generalized to infinite series.) Such a theory—the theory of "generalized soluble groups"—has indeed been created; its foundations were laid by A. G. Kuroš and S. N. Černikov in their paper [8]. We shall use the term *Kuroš–Černikov classes* for the classes of generalized soluble groups occurring within that theory, although some of them had been considered prior to [8]. In this first subsection we give the definitions and indicate the significance of these classes.

The concept basic to the theory of Kuroš–Černikov classes is the following generalization of the idea of a subnormal series. A collection $\mathfrak{S}$ of subgroups of a group G is called a *subnormal system* for G, if it:

(i) contains both 1 and G;

(ii) is fully ordered with respect to inclusion; i.e. for every $A, B \in \mathfrak{S}$, either $A \leq B$ or $B \leq A$;

(iii) is closed under arbitrary unions and intersections; in particular, if for each A in $\mathfrak{S}$, $A \neq G$, we denote by $A^{\#}$ the intersection of all $H \in \mathfrak{S}$ such that $H > A$; and for each B in $\mathfrak{S}$, $B \neq 1$, we denote by $B^{\flat}$ the union of all $H \in \mathfrak{S}$ such that $H < B$, then $A^{\#}, B^{\flat} \in \mathfrak{S}$;

(iv) satisfies the condition that $A \trianglelefteq A^{\#}$ for all $A \in \mathfrak{S}$, $A \neq G$.

The factor groups $A^{\#}/A$ are called the *factors* of the subnormal system $\mathfrak{S}$. We shall say that $\mathfrak{S}$ is *well-ordered upwards* if $A^{\#} \neq A$ for all $A \neq G$, and *well-ordered downwards* if $B^{\flat} \neq B$ for all $B \neq 1$. (Show that these expressions mean simply that $\mathfrak{S}$ is well-ordered by $<$ and $>$ respectively.) A subnormal system is *normal* (rather than just subnormal) if every group in the system is normal in the whole group G. If one subnormal system contains another, then the former is called a *refinement* of the latter. Generalizing the definition in §16.1, we call a normal system *central* if its factors are all central; i.e.

$$A^{\#}/A \leq C(G/A) \quad \text{for all } A \in \mathfrak{S}, A \neq G,$$

or, equivalently,

$$[A^{\#}, G] \leq A \quad \text{for all } A \in \mathfrak{S}, A \neq G.$$

Finally, we shall say that a subnormal system is *soluble* if its factors are all abelian.

We can now define the *Kuroš–Černikov classes*. (The symbols on the left stand for the classes (or properties), while the qualification or property a group must have to be in the class, is stated on the right; we adhere to the traditional symbolism although we do not think it entirely apt.)

SN: the group has a soluble subnormal system of subgroups;

SN*: the group has a soluble subnormal system which is well-ordered upwards;

$\overline{\text{SN}}$: every subnormal system of the group can be refined to a soluble subnormal system;

SI: the group has a soluble normal system;

SI*: the group has a soluble normal system which is well-ordered upwards;

$\overline{\text{SI}}$: every normal system of the group can be refined to a soluble normal system;

Z: the group has a central system;

ZA: the group has a central system which is well-ordered upwards;

ZD: the group has a central system which is well-ordered downwards;

$\bar{Z}$: every normal system of the group can be refined to a central system;

$\tilde{N}$: each subgroup of the group is a member of some subnormal system;

N: each subgroup of the group is a member of some subnormal system which is well-ordered upwards.

22.1.1. Exercise. The condition N is equivalent to the normalizer condition (see §18.2).

22.1.2. Exercise. For finite groups, conditions SN, SN*, $\overline{SN}$, SI, SI*, $\overline{SI}$ are each equivalent to solubility, while conditions Z, ZA, ZD, $\bar{Z}$, $\tilde{N}$, N are equivalent to nilpotency.

22.1.3. Exercise. A group satisfies $\overline{SI}$ (resp. $\bar{Z}$) if and only if all its homomorphic images satisfy SI (resp. Z).

22.1.4. Exercise. Each of the properties SN*, SI*, ZA, $\overline{SN}$, $\overline{SI}$, $\bar{Z}$, $\tilde{N}$, N is preserved by epimorphisms.

The Kuroš–Černikov classes turn out to have very wide compass: for instance, by Magnus' theorem (14.4.4), the class Z includes all free groups. Not so immediate is the assertion that some group from the class $\bar{Z}$ contains as a subgroup a free group of rank >1 (and therefore every countable free group): this is shown by one of the examples (Example 22.2.2) to which we now address ourselves.

22.2. Example

Orderable groups afford important examples of generalized soluble groups.

22.2.1. EXAMPLE. *Every orderable group satisfies* SN. (Recall that a group is *orderable* if it can be given a full order $\leq$ preserved by left and right multiplication, i.e. $a \leq b$ implies $ac \leq bc$ and $ca \leq cb$, for all elements a, b, c; see also §24.2 in the Appendix.) Here is the proof. Let G be an orderable group and let $\leq$ be a full order on it, preserved by left and right multiplication. A subset M of G is called *convex* if together with every pair a, b of elements of M, all elements intermediate between a and b also belong to M; i.e.

$$a, b \in M, \quad a \leq x \leq b \quad \Rightarrow \quad x \in M.$$

We shall show that the collection $\mathfrak{S}$ of all convex subgroups of the ordered group G is a soluble subnormal system. First we verify in order the properties (i) to (iv) defining a subnormal system. Property (i) is obviously possessed by $\mathfrak{S}$. To check property (ii) let A, B be convex subgroups and suppose there is an element $b \in B$, $b \notin A$. We may suppose $b > 1$, since if

$b > 1$ then b^{-1} will serve instead. Clearly we must have $a \leq b$ for all $a \in A$, since otherwise b would be sandwiched between 1 and an element of A, and so would lie in A. Hence for all $a \in A$ either $1 \leq a \leq b$ or $1 \leq a^{-1} \leq b$, whence $A \leq B$. Property (iii) is obvious. To verify (iv) we need to show that whenever A, B are convex subgroups such that $A < B$ and there are no convex subgroups strictly between them, then $A \triangleleft B$. For each $b \in B$ we have $A^b < B^b$ and it is clear that again A^b and B^b are convex subgroups with no convex subgroups strictly between them. Since $B^b = B$ and $\mathfrak{S}$ is fully ordered (by property (ii)), we deduce that $A = A^b$ as required. Finally to see that each factor $A^\#/A$ is abelian, we note that $A^\#/A$ is an ordered group (under the order induced from $\leq$) with no proper convex subgroups. By the classical theorem of Hölder on ordered groups (see for example [3] or [23], where the proof, starting from the definitions, occupies only about a page), such a group is isomorphic to a subgroup of the additive reals, and so in particular is abelian. This completes the proof.

We now turn to examples of generalized nilpotent groups. One important type of such example is afforded by the various congruence subgroups; that is by the classical $\Gamma_n(m)$ over $\mathbf{Z}$, as well as their analogues over other rings. Along these lines we consider the following

22.2.2. EXAMPLE. [Ju. J. Merzljakov, On the theory of generalized soluble and generalized nilpotent groups, Algebra i logika **2**, No. 5 (1963), 29–36.] Let k be the set of all rational numbers (including zero) which as reduced fractions have odd denominator. Clearly k is a ring (obtained from $\mathbf{Z}$ by "localizing at the prime 2"). For $n = 1, 2, \ldots$, define G_n to be the set of all matrices of the form

$$\begin{pmatrix} 1 + 2^n\alpha & 2^n\beta \\ 2^n\gamma & 1 + 2^n\delta \end{pmatrix}, \qquad \alpha, \beta, \gamma, \delta \in k.$$

It is easy to see that each G_n is a group. As we know (from Theorem 14.2.1) the group G_1 (in fact each of the G_n) contains a subgroup free of rank 2, and therefore free subgroups of all ranks up to and including countably infinite. We shall now show that G_1 satisfies the condition $\bar{\mathbf{Z}}$; in view of Exercise 22.1.3 this is equivalent to showing that for every normal subgroup H of G_1, the quotient G_1/H has a central system. We now embark on the proof of this.

First of all, if H is contained in the center of G_1, then, as it is not difficult to verify directly (see the proof of Magnus' theorem 14.4.4), the system of factor groups

$$G_1/H \geq G_2H/H \geq \cdots \geq \bigcap_{n=1}^{\infty} (G_nH/H) \geq 1$$

will be a central system for the factor group G_1/H. Thus we may, and shall, suppose that the normal subgroup H is not contained in the center of G_1. We shall prove that then

$$H \geq T \cap G_n \quad \text{for some } n, \tag{1}$$

where T is the set of matrices in G_n with determinant 1. This will complete the proof, since then G_1/H will be a homomorphic image of $G_1/(T \cap G_n)$, which by Remak's theorem (4.3.9) embeds in the direct product of the abelian group G_1/T and the nilpotent group G_1/G_n.

For the proof of (1) we apply again the fruitful philosophical procedure of "drawing out the chain" (see p. 78).

(i) We first show that H contains a matrix h such that $(h_{11})^2 \neq (h_{22})^2$. Thus let c be any non-scalar matrix in H. It is then easy to verify that if $a = t_{12}(2)$, $b = t_{21}(2)$, then one of the following elements will serve as the element h:

$$c, c^a, c^b, c^a c, c^b c, c^a c^b.$$

(ii) From the existence of the element h above, we get that H contains all transvections $t_{12}(\lambda)$, $t_{21}(\lambda)$ where

$$\lambda = 2\mu((h_{22}/h_{11})^2 - 1),$$

and μ can be any rational which as a reduced fraction has both numerator and denominator odd. This is chiefly a matter of matrix multiplication: first simply check that

$$t_{12}(\lambda) = [t_{12}(2), h^u h]^v, \quad \text{where } u = \begin{pmatrix} -h_{22} & 0 \\ 0 & h_{11} \end{pmatrix}, \quad v = \begin{pmatrix} 1 & 0 \\ 0 & \mu \end{pmatrix}.$$

Then to get $t_{21}(\lambda)$ take the transpose of both sides of the above equation, noting that h may be left untransposed where it occurs on the right hand side, since λ involves only its diagonal elements. The following formula will prove useful in this computation:

$$x^d = \begin{vmatrix} x_{11} & \dfrac{\beta}{\alpha} x_{12} \\ \dfrac{\alpha}{\beta} x_{21} & x_{22} \end{vmatrix}, \quad \text{where } d = \begin{pmatrix} \alpha & 0 \\ 0 & \beta \end{pmatrix}.$$

(iii) The next step is to show that for some $m \geq 1$, the subgroup H contains all transvections of the form $t_{12}(2^m \nu)$, $t_{21}(2^m \nu)$, $\nu \in k$. This is immediate from (i) and (ii).

(iv) Finally, H contains all matrices

$$x = \begin{pmatrix} 1 + 2^{2m}\alpha & 2^{2m}\beta \\ 2^{2m}\gamma & 1 + 2^{2m}\delta \end{pmatrix}, \quad \alpha, \beta, \gamma, \delta \in k, \quad \det x = 1.$$

This is immediate upon checking that

$$x = t_{21}(2^m \xi) t_{21}(2^m \delta) t_{12}(2^m) t_{21}(2^m \eta), \tag{2}$$

where

$$\xi = \frac{2^m \beta - 1}{x_{22}}, \qquad \eta = \frac{2^m \gamma - \delta}{x_{22}}.$$

With this we may stop pulling out elements, as the proof is complete.

Incidentally our proof implies a positive solution to the "congruence problem" (see p. 30) for the group $\mathbf{SL}_2(k)$. For, if H is a normal subgroup of $\mathbf{SL}_2(k)$ of finite index n, then the nth powers of all transvections in $\mathbf{SL}_2(k)$ lie in H, so that the conclusion of Part (iii) holds (in fact $t_{12}(2^m \nu)$, $t_{21}(2^m \nu) \in H$ for all $\nu \in k$, where 2^m is the largest power of 2 dividing n). Since Part (iv) follows from Part (iii), we have that H is a congruence subgroup, as desired.

Note that in the above no special property of the number 2 was used. With slight changes the argument yields for instance the following fact. Let π be a set consisting of all but finitely many of the primes, and denote by π' the (finite) complementary set. Let $\mathbf{Q}_\pi$ denote the ring of rational numbers which as reduced fractions have their denominators divisible only by primes from π. The promised fact is that the subgroups of finite index in $\mathbf{SL}_2(\mathbf{Q}_\pi)$ are congruence subgroups. To see this let H be normal of index n, and let q denote the product of all the primes in π'. Choose the integer $m \geq 1$ large enough so that at most the mth powers of primes in π' divide n. Then H contains all transvections $t_{ij}(q^m \nu)$, $\nu \in \mathbf{Q}_\pi$. The following analogue of the decomposition (2) then yields us our fact if in it we put $\mu = \mu' = q^m$:

$$x \equiv \begin{pmatrix} 1 + \mu\mu'\alpha & \mu\mu'\beta \\ \mu\mu'\gamma & 1 + \mu\mu'\delta \end{pmatrix} = t_{12}(\mu\xi)t_{21}(\mu'\delta)t_{12}(\mu)t_{21}(\mu'\eta), \qquad (2')$$

where

$$\det x = 1, \qquad \xi = \frac{\mu'\beta - 1}{x_{22}}, \qquad \eta = \frac{\mu\gamma - \delta}{x_{22}}.$$

It follows from Example 22.2.2 that the properties $\overline{\mathrm{SI}}$ and $\bar{Z}$ are not inherited by subgroups, since otherwise all countable free groups would have these properties, and thence, by Exercise 22.1.4, all countable groups.

Continuing along the lines of the above, G. A. Noskov [Sib. matem. ž. **14**, No. 3 (1973), 680–683] showed that the group G_1 also has the property $\overline{\mathrm{SN}}$, so that this property also is not inherited by subgroups.

The conditions of finiteness and generalized solubility in topological groups have been investigated by V. M. Gluškov, V. S. Čarin and others (see the survey [5]).

22.3. The Local Theorem

A family $\{M_i \mid i \in I\}$ of subsets of a set M is called a *local covering* of M if M is the union of the M_i, and for every two subsets M_i, M_j in the family there is a subset M_k containing both M_i and M_j. For example the collection of all finite subsets of a given set is a local covering, as is the collection of all finitely generated subgroups of a given group. We say a group H *has a property* σ *locally* if there is a local covering of the group G consisting of subgroups with the property σ. This conforms with our previous use of this term in the case that σ is inherited by subgroups, since, with this proviso, it is clearly equivalent to the condition that all finitely generated subgroups of G have σ. As examples of local properties we may take local finiteness and local nilpotence, which we encountered in the preceding chapter.

We take the opportunity of mentioning that the class of locally finite groups is *properly* contained in the class of periodic groups. The containment is obvious; the difficult part is the strictness. In fact there exist infinite, finitely generated p-groups for each prime p: one family of such groups was given by E. S. Golod (see Example 18.3.2), and another by S. V. Aljošin [Finite automata and Burnside's problem on periodic groups, Matem. zametki **11**, No. 3 (1972), 319–328]. The stronger result (stronger at least for almost all primes) of P. S. Novikov and S. I. Adjan was mentioned without proof on p. 101. Note also the result of Schur that a periodic matrix group (over a field) is locally finite (see [41]).

Since we are on the subject of locally finite groups it is appropriate to include here the following

22.3.1. Theorem (O. Ju. Šmidt). *An extension G of a locally finite group A by a locally finite group G/A is itself locally finite.*

PROOF. We verify that every finite subset M of G generates a finite subgroup. By hypothesis the quotient $\langle M, A \rangle / A$ is finite. By enlarging M if necessary (but still keeping it finite) we may suppose that it contains the inverses of all its elements, and at least one representative from each coset of A in $\langle M, A \rangle$. This assumed, for each pair $x, y \in M$, define $a_{x,y} \in A$ by

$$xy = \overline{xy} \cdot a_{x,y}$$

by choosing $\overline{xy}$ as a representative in M of the coset $\langle M, A \rangle xy$. By iterating, it follows that any product of elements of M can be expressed as the product of an element of M with a product of various of the $a_{x,y}$. Since the totality of the $a_{x,y}$ generates a finite subgroup, the proof is complete.

We shall say that *the local theorem* is true for a given group-theoretic property (and for the class of groups having that property), if local possession implies global possession of the property; i.e. whenever a group has the property locally, then the group itself has the property. For instance the local theorem is true for the class of abelian groups, but not for the class of finite groups.

Why is it that for certain properties the local theorem holds, while for others it does not? Having asked himself this question, A. I. Mal'cev [On a general method of obtaining local theorems in group theory, Uč. Zap.

Ivanovskogo ped. in-ta **1**, No. 1 (1941), 3–9] came to the conclusion that, since local theorems are not specific to group theory, but can be framed just as naturally for rings, loops, and other algebraic systems, the key to its answer must be sought not in group theory or in any other particular theory, but at the foundations of mathematics. And indeed local theorems for widely differing properties were found to have a common conceptual source, consisting in the following simple observation from formal logic: If a theorem is a consequence of the axioms in a certain list, then it is a consequence of only finitely many of the axioms in the list (assuming axioms and theorem expressed in the same language). Elaborating on this observation, Mal'cev arrived at his theorem on the "compactness of the first-order predicate calculus," from which in turn he deduced the truth of the local theorem for any property expressible in terms of so-called "quasi-universal formulae." Thus the problem of whether the local theorem holds for a given property σ, which until the advent of Mal'cev seemed to require a separate solution for each individual σ, was reduced to a general and purely syntactical question: Can σ be expressed in terms of quasi-universal formulae?

By using a clever ruse, which we shall describe here, Mal'cev showed, in the above-mentioned work of 1941, how to express the properties SN, SI, Z in terms of quasi-universal formulae of a special kind, thereby establishing the local theorem for these properties. Later he returned to this method, gave it its present generality, and in the paper "Model correspondences" [Izv. Akad. Nauk SSSR, ser. matem. **23**, No. 3 (1959), 313–336] at one blow obtained almost all the interesting local theorems of group theory. One might say that the 1941 paper first put the "and" in the phrase "the local theorems of algebra and of logic," while that of 1959 dotted the i and crossed the t's. In addition to this, the first paper became the starting-point of model theory.

Here we shall use Mal'cev's method to prove the local theorem for those of the above Kuroš–Černikov classes for which it holds. A self-contained exposition of the logical part of the method is given in §24 in the Appendix; the reader not familiar with the logical ideas involved will need to read that section before continuing.

22.3.2. Theorem (A. I. Mal'cev). *The local theorem is true for the properties* SN, SI, Z, $\overline{\text{SN}}$, $\overline{\text{SI}}$, $\tilde{\text{Z}}$, $\tilde{\text{N}}$.

PROOF. Using the above-mentioned trick of Mal'cev, we shall first show how to translate statements about subnormal systems into formulae in the predicate calculus. Thus with each subnormal system $\mathfrak{S}$ of subgroups of a group G, we associate a binary predicate P on G as follows: define $P(x, y) = T$ (T for "true") if there is a subgroup in $\mathfrak{S}$ which contains x, but does not contain y; otherwise of course define $P(x, y) = F$ (for "false"). It is clear that P satisfies the following universal axioms (in (1) to (10) below the quantifiers $(\forall x)$, $(\forall y)$ etc., are understood):

$$\rceil P(x, x), \tag{1}$$

$$P(x, y) \wedge P(y, z) \rightarrow P(x, z), \tag{2}$$

$$P(x, z) \wedge \rceil P(y, z) \rightarrow P(x, y), \tag{3}$$

$$P(x, z) \wedge P(y, z) \rightarrow P(xy^{-1}, z), \tag{4}$$

$$x \neq e \rightarrow P(e, x), \tag{5}$$

$$P(x, y) \rightarrow P(y^{-1}xy, y). \tag{6}$$

For $y \neq e$, define $A_y = \{x \mid x \in G, P(x, y) = T\}$. It is not difficult to verify that

$$A_y = \bigcup_{y \notin A \in \mathfrak{S}} A, \qquad A = \bigcap_{y \notin A} A_y \quad (A \in \mathfrak{S}, A \neq G);$$

that is, $A_y \in \mathfrak{S}$, and each $A \in \mathfrak{S}$, $A \neq G$, is the intersection of certain of the A_y.

Conversely, suppose we are given a predicate P on G with properties (1) to (6). The sets A_y are then subgroups, and the totality of them is fully-ordered under inclusion. If we adjoin to this collection of A_y all unions and intersections, as well as the whole group G, we end up with a subnormal system for G.

From the above, especially (1) and (3), it follows easily that these processes of going from $\mathfrak{S}$ to P, and from P to $\mathfrak{S}$, are mutually inverse, so that we do indeed have a valid method for translating statements about subnormal systems into formulae in the language of the predicate calculus (PC for short).

The concepts basic to the theory of the Kuroš–Černikov classes translate into the language PC as follows:

solubility of the system: $x \neq e \wedge \rceil P(x, y) \wedge \rceil P(y, x) \rightarrow P([x, y], x),$ (7)

normality of the system: $P(x, y) \rightarrow P(z^{-1}xz, y),$ (8)

centrality of the system: $x \neq e \rightarrow P([x, y], x).$ (9)

We shall also characterize "three-member" systems $1 \leq A \leq G$ (not necessarily subnormal) in terms of the language PC. It is clear that such systems are axiomatized by the formulae (1) to (5) together with the following one:

three-member property: $x \neq e \wedge P(x, y) \rightarrow \rceil P(y, z).$ (10)

It is now easy to see that the properties SN, SI, Z translate as follows:

SN: $(\exists P)(\forall x)(\forall y)(\forall z)((1) \wedge (2) \wedge (3) \wedge (4) \wedge (5) \wedge (7)),$

SI: $(\exists P)(\forall x)(\forall y)(\forall z)((1) \wedge (2) \wedge (3) \wedge (4) \wedge (5) \wedge (7) \wedge (8)),$

Z: $(\exists P)(\forall x)(\forall y)(\forall z)((1) \wedge (2) \wedge (3) \wedge (4) \wedge (5) \wedge (9)),$

where (1), (2), etc. stand for the formulae so numbered; while the properties $\overline{SN}$, $\overline{SI}$, $\overline{Z}$, $\tilde{N}$, all involving refinability, become

$\overline{SN}$: $(\forall P)((P$ subnormal)

$\to (\exists Q)((Q$ soluble subnormal) $\wedge (\forall u)(\forall v)(P(u, v) \to Q(u, v))))$,

$\overline{SI}$: $(\forall P)((P$ normal)

$\to (\exists Q)((Q$ soluble normal) $\wedge (\forall u)(\forall v)(P(u, v) \to Q(u, v))))$,

$\overline{Z}$: $(\forall P)((P$ normal)

$\to (\exists Q)((Q$ central) $\wedge (\forall u)(\forall v)(P(u, v) \to Q(u, v))))$,

$\tilde{N}$: $(\forall P)((P$ three-member)

$\to (\exists Q)((Q$ subnormal) $\wedge (\forall u)(\forall v)(P(u, v) \to Q(u, v))))$,

where of course the expressions (P subnormal) and the like, are meant to be replaced by the appropriate combinations of the formulae (1) to (10). Since, as we now see, all seven properties translate into quasi-universal formulae, we have only to appeal to Theorem 24.3.3 of the Appendix, to obtain the desired conclusion.

For the properties SN*, SI*, ZA, ZD, N, which do not come under the umbrella of Mal'cev's theorem, the local theorem is false: see, in particular, Example 18.2.2.

Note that the formulae for the properties SN, SI, Z are object-universal, so that by Theorem 24.3.1 of the Appendix, these properties are inherited by subgroups. (It is also not difficult to prove this directly.)

22.3.3. Exercise. The properties SN*, SI*, ZA, ZD, N, $\tilde{N}$ are also inherited by subgroups.

As an application of Mal'cev's theorem we use it to generalize Černikov's theorem (19.3.2) to SN-groups.

22.3.4. Theorem (S. N. Černikov). *Every SN-group satisfying the minimal condition is a soluble Černikov group.*

PROOF. The minimal condition immediately gives that G is a periodic SN*-group. It then follows by induction and Šmidt's theorem (22.3.1) that G is locally finite. Since a finite soluble group is an SI-group, and the local theorem is true for the property SI, we deduce that G is an SI-group, and therefore, again by virtue of the minimal condition, an SI*-group. Now we imitate the proof of Theorem 19.3.2, taking into account that the property SI* is inherited by subgroups and homomorphic images. The first two paragraphs of that proof carry over without change, except that we take A to be $1^{\#}$ in an appropriate subnormal system for the SI*-group H. Thus, as in the second paragraph of the proof, we conclude that A is contained in the centre of H, which is therefore nontrivial. Taking the quotient of G by the center $C(H)$ of H and repeating the argument, we deduce that $H/C(H)$ also has nontrivial center. Continuing in this way we discover that the upper central series of H, continued through infinite ordinals (defining $\zeta_\alpha H = \bigcup_{\beta < \alpha} \zeta_\beta H$ for limit ordinals α), is always increasing, so that eventually it must reach H; i.e. H is a ZA-group.

Now let B be a maximal abelian normal subgroup of H. It follows (again as in the second paragraph of the proof of Theorem 19.3.2) that B is contained in the centre of H; however by Theorem 16.2.6 (whose proof is still valid for ZA-groups, although the theorem is stated only for the nilpotent case), B is its own centralizer in H. Hence $B = H$; i.e. H is abelian. This completes the proof.

Appendix: Auxiliary Results from Algebra, Logic and Number Theory

§23. On Nilpotent Algebras

The connexion between nilpotency in groups and nilpotency in rings, which was mentioned in passing in the text proper (§16.1) has indeed an important role to play, enriching as it does both group theory and ring theory. In this section we bring together certain facts about nilpotent algebras, needed for our treatment of Engel groups in §18.3. In the first subsection we give basic definitions, and study the behavior of nilpotency under the standard process of going from associative algebra to Lie algebra, and back. The second subsection is devoted to the construction of non-nilpotent nilalgebras.

23.1. Nilpotency of Associative and Lie Algebras

Let k be a field, and let A be a ring (without any assumption of commutativity or associativity of its multiplicative operation). The ring A is called a *(linear) algebra over* k, if for each $\alpha \in k$, $a \in A$ there is defined an element $\alpha a \in A$, called the "scalar multiple" of a by α, satisfying the following conditions:

$$(\alpha + \beta)a = \alpha a + \beta a,$$

$$(\alpha\beta)a = \alpha(\beta a),$$

$$1a = a,$$

$$\alpha(a + b) = \alpha a + \alpha b,$$

$$\alpha(ab) = (\alpha a)b = a(\alpha b),$$

where α, β are arbitrary elements from k, 1 is the multiplicative identity of k, and a, b are arbitrary elements of A. The algebra A is said to be *associative* if

$$(ab)c = a(bc) \quad \text{for all } a, b, c \in A.$$

The algebra A is called a *Lie algebra* if

$$ab + ba = 0,$$

$$(ab)c + (bc)a + (ca)b = 0,$$

for all a, b, $c \in A$.

The reader will not need to have the role of associativity explained to him; on the other hand the concept of a Lie algebra may seem at first glance to be rather artificial. However the truth is that a Lie algebra is just as natural an object as an associative algebra, since every associative algebra comes accompanied by a Lie algebra, obtained from it in the following standard manner. If A is the associative algebra, and a, b are two of its elements, then we call the element $(a, b) = ab - ba$ their *(ring) commutator*. As for groups one might say that this commutator provides a measure of the departure of the pair of elements from being commutative (cf. §3.2, Ch. 1). It is easy to check that the set A together with the operations of addition, commutation, and scalar multiplication (by scalars from k) is a Lie algebra. This Lie algebra is said to be the Lie algebra *associated* with the associative algebra A. Another source of Lie algebras, namely Lie groups, was mentioned briefly in the main text.

Suppose now that in our associative algebra A we have a vector subspace L which is closed under commutation (so that L is a Lie subalgebra of the Lie algebra associated with A). The associative subalgebra $\hat{L}$ generated in the associative algebra A by the set L, is called the *enveloping algebra* of the Lie algebra L.

An element a of an algebra A is said to be *nilpotent* (of "nil potency" so to speak), if $a^n = 0$ for some integer $n > 0$ (in general depending on a), and for any arrangement of brackets in a^n. The whole algebra A is termed *nilpotent* if there is an integer $n > 0$ such that $a_1 a_2 \cdots a_n = 0$ for all $a_1, a_2, \ldots, a_n \in A$ and for all arrangements of brackets in the left-hand side. The smallest such n is called the *nilpotency class* of A. As mentioned in the main text, the algebra of all (upper) triangular matrices with zero main diagonals is an example of a nilpotent algebra (in this case associative).

23.1.1. Exercise. A finitely generated nilpotent algebra has finite dimension.

23.1.2. Exercise. A Lie algebra L is nilpotent precisely if there exists an $n > 0$ such that $(\cdots((a_1 a_2)a_3)\cdots)a_n = 0$ for all $a_1, \cdots, a_n \in L$.

What happens to nilpotency under the processes of going from an associative algebra to its associated Lie algebra, and in the other direction, that is from the Lie algebra to its enveloping algebra? One half of this question is easily answered: it is almost obvious that if an associative algebra A is nilpotent of class n, then its associated Lie algebra is nilpotent of class $\leq n$ (the inequality not being replaceable, in general, by equality). However, going the other way, we cannot expect that the enveloping algebra of a nilpotent Lie algebra will always be again nilpotent, since there exist non-nilpotent associative algebras whose associated Lie algebras are nilpotent: such, for example, is the algebra A of all triangular matrices with constant main diagonal entries (i.e. constant within each matrix). As another example we may take the direct sum of infinitely many one-generator (or "monogenic") nilpotent algebras, the set of whose classes is unbounded.

The reason for the non-nilpotency of the algebra matrix A just mentioned is strikingly clear: it even has *elements* which are not nilpotent in the associative sense. It turns out that if we demand that the individual elements be nilpotent in the associative sense, and also that the Lie algebra which we are "enveloping," be finitely generated, then nilpotency is preserved under the enveloping process:

23.1.3. Theorem. *Let A be an associative algebra and L a subspace closed under commutation. If L is finitely generated and nilpotent as a Lie algebra, and each of its elements is nilpotent in the associative sense, then the associative enveloping algebra $\hat{L}$ is also nilpotent.*

PROOF. Let $e_1, \ldots, e_s$ be generators of the Lie algebra L. We shall call these elements *generating commutators of weight* 1, and, inductively, if c, d are already defined having weights u, v respectively, then (c, d) is defined to be a *generating commutator of weight $u + v$* (so that an element may have several weights depending on how it is expressed as a commutator). Denote by n the nilpotency class of the Lie algebra L; thus commutators of weight $\geq n$ vanish, so that there are altogether only finitely many nonzero generating commutators, say r of them. Denote them by $c_1, c_2, \ldots, c_r$, where here they are ordered in any manner except that the weights are non-decreasing. By hypothesis, for each i there is a positive integer n_i such that $c_i^{n_i} = 0$. We shall show that the associative algebra $\hat{L}$ is nilpotent of class at most $N = n(n_1 + \cdots + n_r)$. For this it clearly suffices to verify that every product $c_{i_1} \cdots c_{i_N}$ vanishes.

For the latter verification we need a little more terminology. We define the *length* of an expression or product $c_{i_1} \cdots c_{i_m}$ to be m, its *weight* to be the sum of the weights of the factors, an *order inversion* in it to be a segment $c_i \cdots c_j$ with $i > j$, and finally its *characteristic* to be the pair (m, t) where t is the total number of order inversions occurring in it. We order the set of characteristics of such expressions lexicographically; that is we set $(m, t) < (m', t')$, if $m < m'$ or both $m = m'$ and $t < t'$. This is clearly a well-order of the set of characteristics.

Now if there occurs in a product $c_{i_1} \cdots c_{i_m}$ of weight w, an order inversion $c_i c_j$, then upon replacing that segment by $c_j c_i + (c_i, c_j)$ we obtain the sum of two products of the same weight w, but with smaller characteristics. It follows that after a finite number of such replacements we shall end up with a sum of products

$$c_{j_1}^{m_1} \cdots c_{j_l}^{m_l}, \quad \text{where } j_1 < \cdots < j_l, \quad \text{all } m_\alpha \geq 1.$$

It only remains to observe that if $w \geq N$ then each such product vanishes, since then $m_\alpha \geq n_{j_\alpha}$ for at least one m_α, whence $c_{j_\alpha}^{m_\alpha} = 0$. Hence the original product $c_{i_1} \cdots c_{i_m} = 0$ if $m \geq N$, and the proof is complete.

We conclude this subsection with a result about finite generation in rings reminiscent of Exercise 14.3.2 about finite generation in groups. Although the concept of nilpotency does not enter into it, this result is nonetheless an important tool in the study of nilpotency, on which ground we may justify including it here. However, we have a different use for it in mind (see §21.3).

Recall that an additive abelian group A is called a *left module over a ring* k, if for each $\alpha \in k$, $a \in A$, there is defined an element $\alpha a \in A$ in such a way as to satisfy the first four of the axioms listed at the beginning of this subsection (in the definition of an algebra). A *right module* over k is defined analogously. If A has defined on it the structures of both a left and right module over k in such a way that $(\alpha a)\beta = \alpha(a\beta)$ for all $\alpha, \beta \in k$, $a \in A$, then we speak of A as a *bimodule* over k. The *annihilator* of the bimodule A over the ring k is the set of all $\alpha \in k$ such that

$$\alpha a = 0 = a\alpha \quad \text{for all } a \in A.$$

Let K be an associative ring (but not necessarily commutative). If M is a subset of K and s is a positive integer, we write M^s for the subgroup of the additive group of K generated by the set of all products of s elements from M.

23.1.4. Theorem. *Let I be an ideal of the finitely generated associative ring K. If the additive group of the quotient ring K/I is finitely generated, then the ideal I is finitely generated (as ideal). The additive groups of the rings K/I^s, $s = 1, 2, \ldots$, are also finitely generated.*

PROOF. (i) Choose a finitely generated subgroup A of the additive group of the ring K, satisfying the following two conditions:

$$K = A + I; \quad \text{the ring } K \text{ is generated by } A. \tag{1}$$

Let I' be the ideal of K generated by the set $I \cap (A + A^2)$. The sum $A + I'$ is then a subring: for if $a_1, a_2 \in A$, then from the first statement in (1) we get $a_1 a_2 = a + i$ for some $a \in A$, $i \in I$; but then clearly $i \in I'$, whence $a_1 a_2 \in A + I'$.

It then follows from the second statement in (1) that $K = A + I'$. Since A is finitely generated we get that the additive group of K/I' is finitely generated, whence its subgroup I/I' is also finitely generated. It can be seen from its definition that the ideal I' is finitely generated as ideal, since $A + A^2$ is a finitely generated additive group. Hence the ideal I is finitely generated.

(ii) The ring K acts on the additive group of I/I^2 by left and right multiplication, turning it into a K-bimodule. It is clear that I is contained in the annihilator of this bimodule, so that we may consider I/I^2 rather as a bimodule over the quotient ring K/I, in fact, by Part (i), as a *finitely generated* K/I-bimodule. This and the fact that the additive group of K/I is finitely generated, together give us that the additive group of I/I^2 is finitely generated. Hence the additive group of K/I^2 is finitely generated. By an easy induction, of which this is the first step, we deduce that for all $n \geq 0$, the additive group of K/I^{2^n} is finitely generated. The proof is then completed by the trivial observation that for any positive integer s there exists an n such that $2^n > s$, so that the additive groups of all K/I^s are finitely generated.

23.2. Non-Nilpotent Nilalgebras

From now on all algebras will be assumed to be associative. If every element of an algebra is nilpotent, we call it a *nilalgebra*. One of the central questions in the theory of nilpotent algebras consists in the following: Is a finitely generated, associative nilalgebra necessarily nilpotent? For finite-dimensional algebras the answer has long been known to be in the affirmative; however, as E. S. Golod has shown, the answer to the question as it stands is negative. To be more specific, Golod has constructed, for each field k and each integer $d \geq 2$, a non-nilpotent algebra over k on d generators, in which every $(d-1)$-generator subalgebra is nilpotent [On nilalgebras and residually finite p-groups, Izv. Akad. Nauk SSSR, ser. matem. **28**, No. 2 (1964), 273–276]. Golod's construction leads to the resolution of several important group-theoretical questions. The aim of this subsection is to describe that construction.

Let $F = k\{x_1, \ldots, x_d\}$ be the ring of polynomials over a field k in the *non*-commuting variables $x_1, \ldots, x_d$. As a vector space, F has the obvious direct sum decomposition

$$F = F_0 \oplus F_1 \oplus \cdots,$$

where $F_0 \approx k$, and F_n is the subspace spanned by the d^n monomials $x_{i_1} x_{i_2} \cdots x_{i_n}$. The polynomials in F_n will be called *homogeneous of degree n*. An ideal I of the ring F is then said to be *homogeneous* if it is generated by homogeneous polynomials (possibly of various degrees).

23.2.1. Exercise. An ideal I of F is homogeneous if and only if it contains the homogeneous components of each of its elements; i.e. if whenever

$$a \in I, \qquad a = a_{n_1} + \cdots + a_{n_s}, \qquad a_{n_i} \in F_{n_i}, \qquad n_1 < \cdots < n_s,$$

then all $a_{n_i} \in I$.

Let $f_1, f_2, \ldots$ be homogeneous polynomials in F of degrees ≥ 2, ordered so that the degrees are non-decreasing, and suppose also that in this sequence the number r_n of polynomials of degree n is finite for each n. Let I be the ideal generated by $f_1, f_2, \ldots$, and write $A = F/I$. By Exercise 23.2.1

$$A = A_0 \oplus A_1 \oplus \cdots, \quad \text{where } A_n = (F_n + I)/I.$$

In particular, $A_0 \approx F_0 \approx k$, $A_1 \approx F_1$, since the polynomials f_i have degree ≥ 2. The factor algebra F/I, for a certain such ideal I, plays the leading role in the construction we wish to carry out. By way of preparing for the actual construction, we establish a criterion for the infinite-dimensionality of F/I.

Consider the formal power series

$$\sigma = 1 - dt + \sum_{n=2}^{\infty} s_n t^n,$$

where the s_n are integers. It is easy to see that the inverse of this series (in the ring of formal power series over the integers) is the series

$$\sigma^{-1} = \sum_{n=0}^{\infty} y_n t^n,$$

where $y_0 = 1$, $y_1 = d$, and

$$y_n = d y_{n-1} - \sum_{i=2}^{n} s_i y_{n-i} \quad \text{for } n \geq 2.$$

23.2.2. Theorem (E. S. Golod–I. R. Šafarevič). *Let $F = k\{x_1, \ldots, x_d\}$ be the polynomial ring over k in the non-commuting variables $x_1, \ldots, x_d$. Let $f_1, f_2, \ldots$ be homogeneous polynomials from $\dot{F}$ of degree ≥ 2, arranged in non-decreasing order of degree, and let I be the ideal they generate. If the number r_n of polynomials of degree n among the f_i, does not exceed s_n, where the numbers s_n are such that all the coefficients in the series σ^{-1} are non-negative, then the algebra F/I is infinite-dimensional.*

PROOF. (i) Let a_n denote the dimension of the factor algebra $A_n = (F_n + I)/I$ as a vector space over k. What we want to prove is equivalent to the assertion that the series

$$\alpha = \sum_{n=0}^{\infty} a_n t^n$$

is not a polynomial. Put

$$\rho = 1 - dt + \sum_{n=2}^{\infty} r_n t^n,$$

and suppose we already know that the series ρ^{-1} and $\rho\alpha$ have no negative coefficients (we shall prove this in Parts (ii) and (iii) below). Since $\rho^{-1}\rho = 1$, it follows that

$$\rho^{-1}\left(1+\sum_{n=2}^{\infty} r_n t^n\right) = 1 + \rho^{-1} \, dt,$$

and from this and the fact (to be established) that ρ^{-1} has no negative coefficients, it is clear that ρ^{-1} cannot be a polynomial. But then neither can $\alpha = \rho^{-1} \cdot \rho\alpha$ be a polynomial.

(ii) We shall prove that the series ρ^{-1} has no negative coefficients. Working in the ring of formal power series over the integers (where the units are the power series with constant term ± 1) we have

$$\rho = \sigma - \delta = \sigma(1 - \sigma^{-1}\delta), \quad \text{where } \delta = \sum_{n=2}^{\infty} (s_n - \tau_n)t^n,$$

whence

$$\rho^{-1} = \sigma^{-1}(1 - \sigma^{-1}\delta)^{-1} = \sigma^{-1}\left(1 + \sum_{m=1}^{\infty} (\sigma^{-1}\delta)^m\right).$$

Since by hypothesis the coefficients in the series α^{-1} and δ are all non-negative, we deduce the same for ρ^{-1}, as required.

(iii) Finally we prove that the series $\rho\alpha$ has no negative coefficients; i.e. that

$$a_n \geq da_{n-1} - \sum_{i=2}^{n} r_i a_{n-i} \quad \text{for } n \geq 2. \tag{2}$$

Denote by I_n the set of all homogeneous polynomials of degree n in I, and by A_n^* a direct complement of the subspace I_n in the space F_n; i.e. $F_n = I_n \oplus A_n^*$. A comparison of dimensions gives

$$d^n = \dim I_n + a_n.$$

Further, let H_m be the subspace spanned by the polynomials of degree m among $f_1, f_2, \ldots,$ and $\langle XY \rangle$ the subspace spanned by all elements xy, $x \in X$, $y \in Y$ (for arbitrary subsets $X, Y \subseteq F$). We shall prove that

$$I_n \subseteq \langle I_{n-1}F_1 \rangle + \sum_{m=2}^{n} \langle A_{n-m}^* H_m \rangle \quad \text{for } n \geq 2. \tag{3}$$

The desired inequality (2) will then follow by replacing the two sides of (3) by their dimensions, and the symbol $\subseteq$ by the symbol $\leq$.

We thus begin the assault on (3). Let $u \in I_n$. By the definition of I_n, u is a sum of products of the form $vf_j w$ where v, w are homogeneous polynomials. Thus it suffices to show that u is in the right-hand side of (3) in the case $u = vf_j w$. In this case, if w has degree ≥ 1, then clearly $u \in I_{n-1}F_1$. If on the other hand w has degree 0, then we may assume that $u = vf_j$. Suppose that the degree of f_j is m, so that it belongs to the subspace H_m. Then we must have $v \in F_{n-m}$, from which we get, using the direct decomposition of F_{n-m}

above,

$$v = v' + v'', \quad \text{where } v' \in I_{n-m}, \ v'' \in A^*_{n-m}.$$

Hence $u \in \langle I_{n-m}H_m \rangle + \langle A^*_{n-m}H_m \rangle$. Since, clearly, $\langle I_{n-m}H_m \rangle \subseteq \langle I_{n-1}F_1 \rangle$, we get that u is in the right-hand side of (3), as required.

23.2.3. EXAMPLE. *If, in the context of Theorem 23.2.2, we have*

$$r_n \le \varepsilon^2 (d - 2\varepsilon)^{n-2}, \quad \text{where } \varepsilon > 0,$$

then the algebra F/I is infinite-dimensional. By Theorem 23.2.2, to conclude this it suffices to verify that the power series σ^{-1} has all its coefficients nonnegative, where

$$\sigma = 1 - dt + \sum_{n=2}^{\infty} \varepsilon^2 (d - 2\varepsilon)^{n-2} t^n.$$

For this we shall make use of the following familiar formulae for formal power series:

$$\frac{1}{1-a} = \sum_{n=0}^{\infty} a^n; \qquad \frac{1}{(1-a)^2} = \sum_{n=0}^{\infty} (n+1)a^n.$$

Thus

$$\sigma = 1 - dt + \varepsilon^2 t^2 \sum_{n=0}^{\infty} (d - 2\varepsilon)^n t^n$$

$$= 1 - dt + \frac{\varepsilon^2 t^2}{1 - (d - 2\varepsilon)t} = \frac{(1 - (d - \varepsilon)t)^2}{1 - (d - 2\varepsilon)t},$$

whence

$$\sigma^{-1} = \frac{1 - (d - 2\varepsilon)t}{(1 - (d - \varepsilon)t)^2} = (1 - (d - 2\varepsilon)t) \sum_{n=0}^{\infty} (n+1)(d - \varepsilon)^n t^n$$

$$= 1 + \sum_{n=1}^{\infty} (d - \varepsilon)^{n-1} (d + (n-1)\varepsilon) t^n.$$

Since $d - 2\varepsilon \ge 0$, we have $d - \varepsilon \ge \varepsilon > 0$, so that the coefficients of σ^{-1} are in fact all positive.

We are now ready to carry out Golod's construction. The following special case should make the idea of the construction clear.

23.2.4. EXAMPLE. Let k be a field with at most countably many elements, let $F = k\{x, y\}$ be the algebra of polynomials over k in the non-commuting variables x, y, and let F' denote the subalgebra of polynomials with zero constant term. We shall show that in the algebra F there is an ideal I contained within F', such that the algebra $A' = F'/I$, while not itself nilpotent, has all its elements nilpotent (in other words, all its one-generator subalgebras are nilpotent).

Let u_1, u_2, ... be any enumeration of the elements of F'. We begin the definition of a sequence N_1, N_2, ... of positive integers by taking $N_1 = 9$. We then raise u_1 to the power N_1, and decompose the result $u_1^{N_1}$ into the (unique) sum of its homogeneous components $f_1, f_2, \ldots, f_{m_1}$, written in increasing order of degrees. Define N_2 to be any integer exceeding the largest of these degrees, and repeat the process with u_2 and N_2, expressing $u_2^{N_2}$ as the sum of its homogeneous components $f_{m_1+1}, \ldots, f_{m_1+m_2}$, where again the degrees increase with the subscripts. Continuing in this way, we end up with an infinite sequence $f_1, f_2, \ldots$ of polynomials of increasing degrees ≥ 9. Let I be the ideal generated by these polynomials. Since F' is generated by x and y, we have that $A' = F'/I$ is generated by $\hat{x} = x + I$ and $\hat{y} = y + I$, and so is 2-generator. It is also immediate from the way we constructed I that A' is a nilalgebra. It only remains to show that A' is not nilpotent, and for that the infinite-dimensionality of $A = F/I$ suffices (since then $A' = A_1 \oplus A_2 \oplus \cdots$ will also have infinite dimension, and will therefore not be nilpotent: see Exercise 23.1.1 above). In the notation of Theorem 23.2.2, we have $d = 2$, $r_n \leq 1$ (and, in particular, $r_2 = r_3 = \cdots = r_8 = 0$). A direct calculation shows that for $\varepsilon = \frac{1}{4}$ the hypothesis of Example 23.2.3 above is satisfied, so that in this case A is infinite-dimensional as required.

We now give Golod's construction in its full generality.

23.2.5. EXAMPLE. *For each integer $d \geq 2$ and each field k, there exists a non-nilpotent k-algebra on d generators, all of whose $(d-1)$-generator subalgebras are nilpotent.* To begin the proof of this, let $F = k\{x_1, \ldots, x_d\}$ denote, as before, the algebra of polynomials over k in the non-commuting variables $x_1, \ldots, x_d$, and let F' be the subalgebra consisting of all polynomials with zero constant term. We shall construct an ideal I of F, contained within F', such that F'/I has the desired properties.

Let ε be a fixed number, $0 < \varepsilon < \frac{1}{2}$. In view of Example 23.2.3 we may take for I the ideal generated by homogeneous polynomials

$$f_1, \ldots, f_{s_1}, f_{s_1+1}, \ldots, f_{s_2}, \ldots$$

of degree ≥ 2, satisfying condition (4) and the following further condition: for each set of $d - 1$ elements of F' of degree $\leq n$ there should exist a positive integer N such that every product (with repetitions allowed) of N factors drawn from the given $d-1$ elements, lies in the ideal I_n generated by $f_1, \ldots, f_{s_n}$.

We assume that $f_1, \ldots, f_s$, $s = s_{n-1}$, are already defined, and show how to extend the definition to $f_1, \ldots, f_t$, $t = s_n$. Let $g_1, \ldots, g_{d-1}$ be any $d-1$ polynomials of degree $\leq n$ and with zero constant term; we express these polynomials as linear combinations over k of the monomials $M_\alpha(x_1, \ldots, x_d)$ of degree $\leq n$ (with coefficient 1):

$$g_i = \sum_\alpha c_{i\alpha} M_\alpha(x_1, \ldots, x_d), \qquad 1 \leq i \leq d-1, \; c_{i\alpha} \in k.$$

For any positive integer N there are $(d-1)^N$ ways of forming products of N of the g_i's. Regarding now the x_j as constants and the $c_{i\alpha}$ as (commuting) variables, we rewrite each such product as a polynomial in the $c_{i\alpha}$, obtaining as coefficients certain linear combinations $f_{s+1}, \ldots, f_t$ of monomials in the x_j of degrees from among $N, N+1, \ldots, nN$. We shall show that for N sufficiently large the polynomials $f_{s+1}, \ldots, f_t$ have the desired properties. Clearly we have only to satisfy condition (4), since the f's were explicitly constructed to fulfil the other condition. Now condition (4) will hold if we can find an N larger than the degrees of $f_1, \ldots, f_s$, such that the polynomials $f_{s+1}, \ldots, f_t$ number fewer than $\varepsilon^2(d-2\varepsilon)^{N-2}$. We shall show that such an N exists by estimating the size of the set $\{f_{s+1}, \ldots, f_t\}$ in terms of N.

Thus take a typical product $g = g_{i_1} \cdots g_{i_N}$, and suppose that g_i occurs as a factor m_i times. Since the $c_{i\alpha}$ are regarded as *commuting* variables (representing, as they do, elements of the field k), it follows that g, as a polynomial in the $c_{i\alpha}$, has

$$\prod_{i=1}^{d-1} \binom{m_i + q - 1}{q - 1}, \qquad q = d + d^2 + \cdots + d^n,$$

terms. Hence the product g contributes this number of polynomials to the segment $f_{s+1}, \ldots, f_t$ of the list of f's. Since $\binom{u}{v} \leq u^v$ and there are $(d-1)^N$ different g's, it follows that altogether the polynomials $f_{s+1}, \ldots, f_t$ number at most

$$(d-1)^N (N + q - 1)^{(q-1)(d-1)}.$$

Since $d - 1 < d - 2\varepsilon$, this bound will be less than $\varepsilon^2(d-2\varepsilon)^{N-2}$ for N sufficiently large, as required.

§24. Local Theorems of Logic

In this section we shall prove Mal'cev's local theorem for quasi-universal classes. On the assumption that the reader is familiar with the basic concepts of formal logic such as might be given in a general introductory course, our exposition will be fast-paced, though entirely self-contained.

24.1. Algebraic Systems

Let A be a set. A function of n variables defined on A (i.e. with domain A^n) and taking its values in the set $\{T, F\} = \{\text{true, false}\}$ (resp. A) is called an *n-ary predicate* (resp. *n-ary operation*) on A. A set A equipped with predicates $P_\alpha^{n_\alpha}$ and operations $f_\beta^{m_\beta}$ (where n_α, m_β are their "arities") is called an *algebraic system*, and the families of integers n_α, m_β together form its *signature*. Two algebraic systems with the same signature are said to be *isomorphic* if there is a one-to-one correspondence between them (or rather between their underlying sets or "carriers") preserving the predicates and operations. (Thus if A and B are the algebraic systems, with associated $P_\alpha^{n_\alpha}$, $f_\beta^{m_\beta}$ and $Q_\alpha^{n_\alpha}$, $g_\beta^{m_\beta}$ respectively, and $\theta: A \to B$ is the isomorphism, then for instance $P_\alpha^{n_\alpha}$ applied to the n_α-tuple $(a_1, \ldots, a_{n_\alpha})$ of elements of A is to have the same value as $Q_\alpha^{n_\alpha}$ applied to $(a_1\theta, \ldots, a_{n_\alpha}\theta)$.) Groups, rings, fields, vector spaces, partially-ordered sets: these are just a few examples of algebraic systems.

If a subset of an algebraic system is closed under all the operations (as opposed to predicates) of the system then, when taken together with the predicates and operations induced on it (i.e. the restrictions to the subset of the operations and predicates of the original system), the subset is called a *subsystem*.

As mentioned in the main body of the text (§22.3), a family A_i, $i \in I$, of subsets of a set A is said to form a *local covering* of A, if $A = \bigcup_{i \in I} A_i$, and for each pair A_i, A_j there is a third A_k containing them both. The idea of a local covering is closely bound up with that of a filter, which concept we now introduce.

Let I be a set. A collection $\mathscr{F}$ of nonempty sets of I is said to be *closed under finite intersections* if it contains the intersection of each pair of its sets: $X \in \mathscr{F}$, $Y \in \mathscr{F} \Rightarrow X \cap Y \in \mathscr{F}$. The collection $\mathscr{F}$ is called a *filter* on I if in addition to being closed under finite intersections it has the property that it contains all supersets of each of its sets: $X \in \mathscr{F}$, $Y \supseteq X \Rightarrow Y \in \mathscr{F}$. A filter not contained in any larger filter, or in other words a maximal filter, is termed an *ultrafilter*. (Note that the collection of all subsets of I is not a filter since it contains the empty set.) It is easily verified that a filter $\mathscr{F}$ is an ultrafilter if and only if for each subset of I, either the subset itself or its complement lies in $\mathscr{F}$. It is obvious that every collection of nonempty subsets of I which is closed under finite intersections is contained in some filter (just throw in all the supersets of members of the collection), and that, granted Zorn's lemma, every filter is contained in an ultrafilter. We mention in passing the fact that no ultrafilter has yet been constructed without using Zorn's lemma, except for the uninteresting ones consisting of all subsets containing a particular element of I.

Let $(A_i, i \in I)$ be a local covering of a set A. For each $\alpha \in I$ define

$$I_\alpha = \{i \,|\, i \in I,\ A_i \supseteq A_\alpha\}.$$

If A_α, $A_\beta \subseteq A_\gamma$ then clearly $I_\gamma \subseteq I_\alpha \cap I_\beta$. Hence the intersection of any finite number of the I_α is nonempty, and so the family of I_α can be completed to a collection of nonempty subsets of I closed under finite intersections, and thence to an ultrafilter $\mathscr{F}$ containing the I_α. If each A_i comes equipped with a predicate P_i (all P_i being of the same arity), then we denote by $P = \lim P_i$ the predicate on A defined by the following rule:

$$P(x, y, \ldots) = T \text{ on } A \Leftrightarrow \{i \mid x, y, \ldots \in A_i, P_i(x, y, \ldots) = T \text{ on } A_i\} \in \mathscr{F}.$$

It may happen that the restriction of P to some A_i does not coincide with P_i. On the other hand if we take a predicate Q defined on A, take its restriction Q_i to each A_i and then take the limit of the Q_i over the filter $\mathscr{F}$, as above, then we do get Q back again.

24.2. The Language of the Predicate Calculus

The alphabet of the language of the predicate calculus (PC for short) consists of: object variables, predicate variables, brackets, and the following logical symbols (of which the first four are called *connectives*, and the last two *quantifiers*):

$\wedge$	and
$\vee$	or
$\rceil$	not
$\rightarrow$	implies
$=$	equals
$\forall$	for all
$\exists$	there exists

If in a formula (i.e. statement) in the language of the predicate calculus there is no quantification over predicates (as opposed to objects), then we say that it is a formula in the *first-order predicate calculus* (FOPC). Although the language of PC seems rather meager, it is nonetheless quite expressive. For example the usual axioms (or defining properties) for the class of torsion-free abelian groups can be expressed in the language of PC—even FOPC—supplemented by the symbols for the binary, unary and nullary operations $\cdot$, $^{-1}$, e of group theory:

(1) associativity: $(\forall x)(\forall y)(\forall z)((xy)z = x(yz))$,

(2) property of the identity: $(\forall x)(xe = x \wedge ex = x)$,

(3) property of inverse elements: $(\forall x)(x^{-1}x = e \wedge xx^{-1} = e)$,

(4) commutativity: $(\forall x)(\forall y)(xy = yx)$,

(5) torsion-freeness: $(\forall x)(x = e \vee \rceil(\underbrace{x \cdots x}_{n} = e))$, $n = 1, 2, \ldots$.

We now give a few more examples. A binary predicate $\leq$ is called a *partial order* if it has the following properties:

(6)   reflexivity: $(\forall x)(x \leq x)$,

(7)   transitivity: $(\forall x)(\forall y)(\forall z)(x \leq y \wedge y \leq z \rightarrow x \leq z)$,

(8)   anti-symmetry: $(\forall x)(\forall y)(x \leq y \wedge y \leq x \rightarrow x = y)$,

and it is called a *full* (or *linear*) *order* if in addition it satisfies:

(9)   fullness: $(\forall x)(\forall y)(x \leq y \vee y \leq x)$.

A (binary) predicate $\leq$ defined on a group is said to be *preserved by the group operation*, if

(10) $(\forall x)(\forall y)(\forall z)(x \leq y \rightarrow xz \leq yz \wedge zx \leq zy)$.

A group is defined to be *orderable* (or an *O-group*) if there exists a full order on it which is preserved by multiplication. A group is an *O^*-group* if every partial order on it which is preserved by multiplication can be extended to a full order preserved by multiplication. It is easily seen that the following statements (or axioms) in the language of PC (but no longer FOPC) distinguish the classes of orderable and O^*-groups respectively, from the class of all groups:

(11)   orderability: $(\exists P)(\forall x)(\forall y)(\forall z)(P(x, x) \wedge$

$$(P(x, y) \wedge P(y, z) \rightarrow P(x, z)) \wedge$$

$$\wedge (P(x, y) \wedge P(y, x) \rightarrow x = y) \wedge (P(x, y) \vee P(y, x)) \wedge$$

$$\wedge (P(x, y) \rightarrow P(xz, yz) \wedge P(zx, zy))),$$

(12) O^*-property: $(\forall P)((P \text{ is a p.p.o.})$

$$\rightarrow (\exists Q)((Q \text{ is a p.f.o.}) \wedge (\forall u)(\forall v)(P(u, v) \rightarrow Q(u, v))))).$$

In the formula (12) for the sake of brevity we have used the abbreviations p.p.o. and p.f.o. for "preserved partial order" and "preserved full order" respectively; the two statements involving these abbreviations can easily be written in the language of PC by combining (6), (7), (8), (9) and (10).

Any formula of PC can by a systematic sequence of alterations be put in an equivalent *prenex normal form*, in which all quantifiers are grouped at the beginning, before all the other logical symbols. A formula in prenex normal form is said to be *universal* if the existential quantifier $\exists$ does not appear in it, and *object-universal* if there are no occurrences of the quantifier $\exists$ referring to (i.e. immediately followed by) an object variable. A formula of PC is *quasi-universal* if it is obtained from object-universal statements without free object-variables, by first combining them using connectives only, and then using the universal quantifier $\forall$ to bind every predicate-variable. (This definition differs somewhat from Mal'cev's original 1959 definition: the new version was suggested by J. P. Cleave [Local properties of systems, J. London Math. Soc. (1), **44** (1969), 121–130; Addendum, J. London Math. Soc. (2), **1** (1969), 384] who uses instead the term "Boolean-universal.") The formulae (1) to (10) are all examples of universal statements. The formula (11) is object-universal, while (12) can be changed to a quasi-universal formula equivalent to it by extracting all the quantifying symbols (i.e. $(\forall u)$, etc.) from the field of reference of the symbol $(\exists Q)$ and restoring them to positions immediately following $(\exists Q)$, to get the prenex normal form of the formula $(\exists Q)(\cdots)$ in (12).

24.2.1. Exercise. The simple groups can be distinguished within the class of all groups by means of a quasi-universal axiom.

24.3. The Local Theorems

We first consider object-universal formulae.

24.3.1. Theorem. *Let Φ be an object-universal formula. If Φ is true on an algebraic system (i.e. when the object-domain is taken to be the system), then it is true on every subsystem. If Φ is true on subsystems A_i which form a local covering of an algebraic system A, then Φ is true on A.*

PROOF. In order to avoid tedious and obfuscating notational complexities, we shall limit our proof to the case where the algebraic system carries a single binary operation denoted by $\cdot$, and a single binary predicate S.

(i) Let A be an algebraic system with this signature, satisfying the object-universal axiom Φ, and let $\hat{A}$ be a subsystem of A (recall that this means that $\hat{A}$ is closed under the operation $\cdot$, and that its ($\hat{A}$'s) operation and predicate are the restrictions to it of the operation and predicate of the original algebra A. We wish to prove that Φ is true on $\hat{A}$. Suppose first that Φ contains no quantifiers. In this case the desired conclusion is immediate from the definition of what it means for a formula of PC to be true on an algebraic system (namely that the formula remain true whatever values the free object- and predicate-variables are assigned from within the system and from the set of predicates on the system respectively). Suppose next that Φ does contain quantifiers, and use induction on their number. Depending on what sort of quantifier is the last applied, there are three possibilities:

$$\Phi = (\forall t)\Psi; \qquad \Phi = (\forall R)\Psi; \qquad \Phi = (\exists R)\Psi,$$

where t is an object-variable, R a predicate-variable, and Ψ an object-universal formula with one fewer quantifier. By extending predicates on $\hat{A}$ arbitrarily to A (for the second case), restricting predicates on A to $\hat{A}$ (for the third case), and using the inductive hypothesis in all three cases, we conclude that Φ is true on $\hat{A}$, as required.

(ii) Suppose the subsystems A_i, $i \in I$, form a local covering of the algebraic system A, and that the object-universal formula Φ is true on the A_i. We wish to show that then Φ is true on A. Let $x, y, \ldots$ be the free object-variables, and $P, Q, \ldots$ the free predicate-variables contained in Φ; we express this by writing

$$\Phi = \Phi(x, y, \ldots, P, Q, \ldots).$$

Let $\mathcal{F}$ be an ultrafilter on I, constructed from the local covering $(A_i, i \in I)$ as indicated at the end of §24.1 above. For each $i \in I$, let P_i, Q_i be predicates on A_i, and write $P_0 = \lim P_i$, $Q_0 = \lim Q_i$. To get the desired conclusion it clearly suffices to show that, whatever values are given to $x, y, \ldots$ from A, we shall have

$$\{i \,|\, x, y, \ldots \in A_i, \Phi(x, y, \ldots, P_i, Q_i, \ldots) = T \text{ on } A_i\} \in \mathcal{F} \Rightarrow$$
$$\Rightarrow \Phi(x, y, \ldots, P_0, Q_0, \ldots) = T \text{ on } A. \tag{1}$$

We now prove this, again by induction on the number of quantifiers occurring in the formula Φ. Thus to begin with suppose that Φ contains no quantifiers; we shall show that then both (1) and the reverse implication hold. Denote by (1′) the union of these two implications (i.e. (1′) is the double implication). If Φ also contains no connectives, then one of the following three possibilities must occur:

$$\Phi = (u(x, y, \ldots) = v(x, y, \ldots)); \quad \Phi = S(u, v); \quad \Phi = P(u, v, \ldots),$$

where $u, v, \ldots$ are products (with arrangements of brackets), and S and P are as above. In the first two cases (1′) is obvious, while in the third case it follows from the definition of lim (with P_i the restriction of P to A_i). Suppose next that Φ does contain connectives (but still no quantifiers). Since the formula $\Psi \rightarrow \Omega$ is equivalent to $\,]\,\Psi \vee \Omega$, we may suppose that Φ involves only the connectives $\wedge, \vee,]$; we shall then establish (1′) by induction on the number of these connectives in Φ. Depending on which connective is the last used in building up Φ, there are again three possibilities:

$$\Phi = \Psi \wedge \Omega; \quad \Phi = \Psi \vee \Omega; \quad \Phi = \,]\,\Psi.$$

In all three cases (1′) follows easily from the inductive hypothesis and the definition of the filter $\mathcal{F}$.

As promised, we shall prove (1) by induction on the number of quantifiers appearing in (the prenex normal form of) Φ. We have established (1′) when Φ contains no quantifiers; proceeding to the inductive step, suppose now that Φ does contain quantifiers. Once again three possibilities arise, depending on which quantifier applies last in Φ:

$$\Phi = (\forall t)\Psi; \quad \Phi = (\forall R)\Psi; \quad \Phi = (\exists R)\Psi, \tag{2}$$

where t is an object-variable, R a predicate-variable, and Ψ is an object-universal formula with one fewer quantifier.

If the first possibility occurs, we argue as follows. For whatever value t takes in A, we have to show that

$$\Psi(t, x, y, \ldots, P_0, Q_0, \ldots) = T \text{ on } A. \tag{3}$$

Since by inductive hypothesis (1) is true with $\Psi(t, x, y, \ldots)$ in place of Φ, to get (3) it suffices to show that the set

$$\{i \mid t, x, y, \ldots \in A_i, \Psi(t, x, y, \ldots, P_i, Q_i, \ldots) = T \text{ on } A_i\}$$

belongs to $\mathscr{F}$. However this set contains the intersection of the set $\{\cdots\}$ of (1) with the set $\{i \mid t \in A_i\}$, so does indeed belong to $\mathscr{F}$.

We next consider the second possibility in (2). Let R_0 be an arbitrary predicate on A, and R_i its restriction to A_i. As mentioned at the end of §24.1 above, $R_0 = \lim R_i$. We have to show that

$$\Psi(x, y, \ldots, P_0, Q_0, R_0, \ldots) = T \text{ on } A.$$

By the inductive hypothesis, for this it suffices to verify that the set

$$\{i \mid x, y, \ldots \in A_i, \Psi(x, y, \ldots, P_i, Q_i, R_i, \ldots) = T \text{ on } A_i\}$$

belongs to $\mathscr{F}$. However this set contains the set $\{\cdots\}$ in (1), so that this case is also finished.

For the third and last case in (2), suppose again that the left half (i.e. the hypothesis) of the implication (1) is true. Then for each $i \in I$, there is a predicate R_i on A_i such that the set

$$\{i \mid x, y, \ldots \in A_i, \Psi(x, y, \ldots, P_i, Q_i, R_i, \ldots) = T \text{ on } A_i\}$$

belongs to $\mathscr{F}$. Writing $R_0 = \lim R_i$, we have by the inductive hypothesis that

$$\Psi(x, y, \ldots, P_0, Q_0, R_0, \ldots) = T \text{ on } A.$$

However this means that the second half (i.e. the conclusion) of the implication (1) holds. This completes the inductive step and thereby the proof of the theorem.

To illustrate the use of this theorem we point out how the local theorem for the class of orderable groups follows from it: one has merely to observe that the axiom (11) of §24.2 defining that class, is object-universal.

24.3.2. Exercise. If in a group G each finitely generated subgroup contains an abelian normal subgroup of finite index $\leq n$, then the whole group G contains an abelian normal subgroup of index $\leq n$.

We now turn to quasi-universal formulae.

24.3.3. Theorem (A. I. Mal'cev). *If a quasi-universal formula Φ is true on subsystems A_i locally covering an algebraic system A, then Φ is true also on A.*

PROOF. By the definition of quasi-universality, Φ has the form

$$\Phi = (\forall P_1) \cdots (\forall P_n)\hat{\Phi},$$

where $\hat{\Phi}$ is built up by means of the connectives $\wedge$, $\vee$, $]$, $\rightarrow$, from object-universal formulae without free object-variables, but with free predicate-variables from among $P_1, \ldots, P_n$. Suppose that Φ is true on each A_i; we wish to prove that then it is true on A. In other words we wish to show that, given any substitution of predicates on A for $P_1, \ldots, P_n$, and given also that $\hat{\Phi}$ is true on each A_i for the predicates on A_i induced from those substituted for $P_1, \ldots, P_n$, then $\hat{\Phi}$ is true on A. Thus if we regard $P_1, \ldots, P_n$ as predicates on A, and use the same symbols for their restrictions to A_i, then our problem reduces to proving the theorem for $\hat{\Phi}$, rather than Φ. It is a fact usually proved in elementary courses in logic, that $\hat{\Phi}$ can be rewritten in an equivalent form as

$$\hat{\Phi} = \wedge \Phi_\alpha, \qquad \Phi_\alpha = \vee \Phi_{\alpha\beta},$$

where the $\Phi_{\alpha\beta}$ are object-universal formulae containing no free object-variables, or the negations of such formulae. It is clear that the theorem is true for $\hat{\Phi}$ if it is true for all the Φ_α. This and the argument preceding it reduce the problem to the consideration of the following three cases:

$$\Phi = \Psi_1 \vee \cdots \vee \Psi_r;$$

$$\Phi =]\Omega_1 \vee \cdots \vee]\Omega_s;$$

$$\Phi = \Psi_1 \vee \cdots \vee \Psi_r \vee]\Omega_1 \vee \cdots \vee]\Omega_s, \qquad (4)$$

where $r, s \geq 1$, and the Ψ_α, Ω_β are object-universal formulae containing no free object-variables.

We shall consider in detail only the third of these cases. Thus with Φ as in (4), suppose Φ is true on each A_i, but false on A. Then all the Ψ_α are false on A, while the Ω_β are all true on A. It follows from the second part of Theorem 24.3.1 above that therefore each Ψ_α is false on some $A_{i(\alpha)}$. Let A_i be a subsystem containing all $A_{i(\alpha)}$. By the first part of Theorem 24.3.1 each Ψ_α is false on A_i, while each Ω_β is true on A_i. Hence Φ is false on A_i, contrary to hypothesis. The other two cases yield to analogous, though simpler, arguments. This completes the proof.

By way of illustrating how Theorem 24.3.3 can be used, note that it implies the local theorem for the class of O^*-groups (see formula (12) in §24.2), and for the class of simple groups (see Exercise 24.2.1).

§25. On Algebraic Integers

We shall assume in what follows that the reader is familiar with the basic ideas and terminology of field theory, such as might be encountered in a general algebra course.

Let $\hat{\mathbf{Q}}$ denote the algebraic closure of the field $\mathbf{Q}$ of rational numbers. The elements of $\hat{\mathbf{Q}}$ are called *algebraic numbers*; equivalently, an algebraic number is a number (i.e. an element of $\mathbf{C}$) satisfying a polynomial with integer coefficients. From among the algebraic numbers we single out for special attention the *algebraic integers*: these are the algebraic numbers satisfying polynomials with integer coefficients and leading coefficient 1. The term *algebraic number field* is reserved for those subfields of $\hat{\mathbf{Q}}$ of finite degree (i.e. of finite dimension as a vector space) over $\mathbf{Q}$; in other words, an algebraic number field is a "finite extension" of $\mathbf{Q}$.

For the rest of this subsection, k will denote an algebraic number field and k_0 the subset of algebraic integers in k.

25.1.1. Exercise. For each $\alpha \in k$, there is an $m \in \mathbf{Z}$ such that $m\alpha \in k_0$.

25.1.2. Exercise. $\mathbf{Q}_0 = \mathbf{Z}$.

25.1.3. Lemma. *The subset k_0 is a subring of k.*

PROOF. Let $\alpha, \beta \in k_0$, and let γ be any of the elements $\alpha \pm \beta$, $\alpha\beta$. We wish to prove that $\gamma \in k_0$. Since k is a finite extension of $\mathbf{Q}$, there exist positive integers l, m such that:

$$\alpha^l = \sum_{i=0}^{l-1} a_i \alpha^i, \quad a_i \in \mathbf{Z}; \qquad \beta^m = \sum_{j=0}^{m-1} b_j \beta^j, \qquad b_j \in \mathbf{Z}.$$

It is then easily seen that the totality of elements of the form

$$\sum_{i=0}^{l-1} \sum_{j=0}^{m-1} c_{ij} \alpha^i \beta^j, \qquad c_{ij} \in \mathbf{Z},$$

is a subring A of k, whose additive group is finitely generated and (of course) torsion-free. Let $\{\gamma_1, \ldots, \gamma_t\}$ be a basis of this additive group. Since $\gamma \in A$ we have

$$\gamma_r \gamma = \sum_{s=1}^{t} d_{rs} \gamma_s, \qquad d_{rs} \in \mathbf{Z}.$$

But this implies that the characteristic polynomial of the matrix (d_{rs}) has γ as a root. Since the characteristic polynomial has leading coefficient ± 1, it follows that γ is an algebraic integer, and the lemma is proved.

It is clear that for each $\alpha \in k$, the map defined by $x \to x\alpha$ for all $x \in k$, is a linear transformation of the vector space k over the field $\mathbf{Q}$. We shall denote by χ_α the characteristic polynomial of this linear transformation, and by $\mathrm{tr}(\alpha)$ its trace. More generally, if K is a subfield of $\mathbf{C}$, and L is a finite extension of K (i.e. has finite degree over its subfield K), then for each $\alpha \in L$ the map defined by $x \to x\alpha$ for all $x \in L$ is linear over K, and its trace (which of course lies in K) is denoted by $\mathrm{tr}_{L/K}(\alpha)$.

25.1.4. Lemma. *If $\alpha \in k_0$, then χ_α is a polynomial over **Z**.*

PROOF. Let n be the degree of k over **Q**. Fix on a basis for k over **Q**, and then with each $\gamma \in k$ associate the matrix $\tilde{\gamma}$ relative to this basis, of the linear transformation $x \to x\gamma$, $x \in k$. This obviously defines a ring monomorphism $k \to M_n(\mathbf{Q})$. Thus since α satisfies a monic polynomial over **Z**, so will the characteristic roots of the matrix $\tilde{\alpha}$. Now the coefficients of the characteristic polynomial χ_α are, up to sign, sums of products of these characteristic roots, so by Lemma 25.1.3 they are algebraic integers. But then by Exercise 25.1.2 they lie in **Z**, as desired.

25.1.5. Lemma. *If $\{\omega_1, \ldots, \omega_n\}$ is a basis for k over **Q**, then: (i) the matrix $(\mathrm{tr}(\omega_i\omega_j))$ is nonsingular; and (ii) there exists a basis $\{\omega_1^*, \ldots, \omega_n^*\}$ for k over **Q**, dual to the given basis; i.e. satisfying*

$$\mathrm{tr}(\omega_i\omega_j^*) = \delta_{ij}.$$

PROOF. (i) Suppose on the contrary that between the columns of the matrix $(\mathrm{tr}(\omega_i\omega_j))$ there is a nontrivial dependence relation with coefficients $\alpha_j \in \mathbf{Q}$. Then

$$\omega = \sum_{j=1}^{n} \alpha_j\omega_j \neq 0; \qquad \mathrm{tr}(\omega_i\omega) = 0 \quad \text{for } i = 1, \ldots, n.$$

Since $\{\omega_1\omega, \ldots, \omega_n\omega\}$ is also a basis for k over **Q**, we must have

$$1 = \sum_{j=1}^{n} \beta_j\omega_j\omega \quad \text{for suitable } \beta_j \in \mathbf{Q}.$$

Since $n \geq 1$, the trace of the right-hand side of this equation is 0, while the trace of the left-hand side is of course 1. This contradiction completes the proof of (i).

(ii) We look for the ω_j^* in the form

$$\omega_j^* = \sum_{l=1}^{n} \xi_{jl}\omega_l, \qquad \xi_{jl} \in \mathbf{Q}.$$

On substituting from this, the conditions $\mathrm{tr}(\omega_i\omega_j^*) = \delta_{ij}$, $i = 1, \ldots, n$, become a system of linear equations in the unknowns $\xi_{j1}, \ldots, \xi_{jn}$ whose coefficient matrix is just $(\mathrm{tr}(\omega_i\omega_l))$. By (i) this matrix is nonsingular so that the system has a solution. This completes the proof of the lemma.

25.1.6. Theorem. *The additive group of the ring k_0 is finitely generated.*

PROOF. As before let $\{\omega_1, \ldots, \omega_n\}$ be a basis for k over **Q**. In view of Exercise 25.1.1 we may assume that the ω_i lie in k_0. Let $\{\omega_1^*, \ldots, \omega_n^*\}$ be a dual basis. Then, again by Exercise 25.1.1 there exists a non-zero integer m such that $m\omega_j^* \in k_0$, $j = 1, \ldots, n$. Now let γ be an arbitrary element of k_0 and express it in terms of the ω_i:

$$\gamma = \sum_{i=1}^{n} \gamma_i \omega_i, \qquad \gamma_i \in \mathbf{Q}.$$

Multiplying this equation by $m\omega_j^*$ and then taking the traces of both sides of the resulting equation, we obtain, for each $j = 1, \ldots, n$, $m\gamma_j = \mathrm{tr}(m\gamma\omega_j^*)$. Since $m\gamma\omega_j^* \in k_0$, we deduce that $m\gamma_j \in \mathbf{Z}$ (Lemma 25.1.4). Since γ was arbitrary (in k_0) it follows that the additive group of k_0 is contained in the additive group generated by $\omega_1/m, \ldots, \omega_n/m$, and so (by Exercise 8.1.7) is itself also finitely generated, as required.

25.1.7. Lemma. *Let K be a subfield of $\mathbf{C}$, and let L be a finite extension of K. If the degree of L over K is d, then there exist exactly d monomorphisms of L into $\mathbf{C}$ which fix K elementwise. Further, if these monomorphisms are denoted by $\sigma_1, \ldots, \sigma_d$, then*

$$\mathrm{tr}_{L/K}(\alpha) = \alpha\sigma_1 + \cdots + \alpha\sigma_d \quad \text{for each } \alpha \in L. \tag{1}$$

Finally, if $\{\omega_1, \ldots, \omega_d\}$ is a basis for L over K, then

$$\det(\mathrm{tr}_{L/K}(\omega_i\omega_j)) = \det(\omega_i\sigma_j)^2 \tag{2}$$

PROOF. (i) We begin with the first assertion. Let $\alpha \in L$, let f be the minimal polynomial of α over K, and let $\alpha_1, \ldots, \alpha_m$ be the roots of f in $\mathbf{C}$. Clearly there is a one-to-one correspondence between the monomorphisms $K(\alpha) \to \mathbf{C}$ which induce the identity map on K, and the assignments $\alpha \to \alpha_i$ of possible images to α. We might now appeal to the well-known theorem which says that finite extensions of fields of characteristic zero are simple extensions (see, for example, [40], p. 126), to conclude that there exists an element α such that $L = K(\alpha)$, whence the assertion would be immediate. Alternatively, in the present situation we can sidestep that theorem by means of the following device: write L in the form $K(\gamma_1, \ldots, \gamma_s)$ and observe that each monomorphism $K(\gamma_1, \ldots, \gamma_i) \to \mathbf{C}$ which fixes K elementwise, can, as before, be extended in exactly $|K(\gamma_1, \ldots, \gamma_{i+1}) : K(\gamma_1, \ldots, \gamma_i)|$ ways to a monomorphism from $K(\gamma_1, \ldots, \gamma_{i+1})$ to $\mathbf{C}$.

(ii) We next prove (1). Consider first the special case that α is such that $L = K(\alpha)$. Then the minimal polynomial of α over K has degree $|L:K| = d$, and is therefore, up to sign, the characteristic polynomial of the linear transformation $x \to x\alpha$ of the vector space L over the field K. As in Part (i) we have that $\alpha\sigma_1, \ldots, \alpha\sigma_d$ comprise all the roots of the minimal polynomial, and hence also of this characteristic polynomial, whence the equation (1).

The general case reduces to the above special case with the help of the equality

$$\mathrm{tr}_{L/K}(\alpha) = |L:K(\alpha)| \cdot \mathrm{tr}_{K(\alpha)/K}(\alpha),$$

which we now prove. Let $\{\omega_i\}$ be a basis for L over $K(\alpha)$, and $\{\delta_j\}$ a basis for

$K(\alpha)$ over K. Then $\{\omega_i\delta_j\}$ is a basis for L over K, and in terms of this basis the matrix of the linear transformation $x \to x\alpha$ is a block-diagonal matrix (i.e. with the off-diagonal blocks all zero) where the diagonal blocks are just repetitions of the matrix relative to $\{\delta_j\}$ of the transformation $x \to x\alpha$ of the vector space $K(\alpha)$ over K.

(iii) Finally we prove (2). From (1) we get

$$\text{tr}_{L/K}(\omega_i\omega_j) = \sum_{l=1}^{d} (\omega_i\sigma_l) \cdot (\omega_j\sigma_l).$$

Since the right-hand side of this equation is just the (i,j)th entry in the product of the matrix $(\omega_r\sigma_s)$ with its transpose, (2) follows. This concludes the proof of the lemma.

For each monomorphism $\phi: L \to \mathbf{C}$ there is a *conjugate* monomorphism $\bar{\phi}: x \to \overline{x\phi}$, where the bar denotes complex conjugation. If $\phi = \bar{\phi}$ then we say that the monomorphism ϕ is *real*, and in the contrary case, *complex*.

Return now to the field k. Let $n = |k: \mathbf{Q}|$, and let s and $2t$ be respectively the numbers of real and complex monomorphisms $k \to \mathbf{C}$, so that, by Lemma 25.1.7, $n = s + 2t$. Write $\sigma_1, \ldots, \sigma_s$ for the real monomorphisms, and $\sigma_{s+1}, \bar{\sigma}_{s+1}, \ldots, \sigma_{s+t}, \bar{\sigma}_{s+t}$ for the complex ones.

Consider n-dimensional Euclidean space $\mathbf{R}^n$, and the map $\sigma: k \to \mathbf{R}^n$, defined by

$$x\sigma = (x\sigma_1, \ldots, x\sigma_s, \text{Re}(x\sigma_{s+1}), \text{Im}(x\sigma_{s+1}), \ldots$$

$$\ldots, \text{Re}(x\sigma_{s+t}), \text{Im}(x\sigma_{s+t})), \qquad x \in k,$$

where Re, Im denote, as usual, the taking of real and imaginary parts.

25.1.8. Exercise. The map σ is a monomorphism from the additive group k to the additive group $\mathbf{R}^n$.

Next, letting k^* denote, as usual, the multiplicative group of the field k, we define a map $\tau: k^* \to \mathbf{R}^{s+t}$ by

$$x\tau = (\ln|x\sigma_1|, \ldots, \ln|x\sigma_{s+t}|), \qquad x \in k^*.$$

25.1.9. Exercise. The map τ is a homomorphism from k^* to the additive group $\mathbf{R}^{s+t}$.

25.1.10. Lemma. *The set $k_0\sigma$ is discrete in $\mathbf{R}^n$; i.e. it intersects each ball in $\mathbf{R}^n$ in a finite set. If k_0^* denotes the multiplicative group of the ring k_0, then the set $k_0^*\tau$ is discrete in $\mathbf{R}^{s+t}$.*

PROOF. Let $\{\omega_1, \ldots, \omega_n\}$ be a basis for the additive group of the ring k_0. (The fact that this group has rank n is implicit in the proof of Theorem 25.1.6.)

Since $\det(\mathrm{tr}(\omega_i\omega_j)) \neq 0$ (Lemma 25.1.5), the n-tuples

$$(\omega_i\sigma_1, \ldots, \omega_i\sigma_s, \omega_i\sigma_{s+1}, \omega_i\bar{\sigma}_{s+1}, \ldots, \omega_i\sigma_{s+t}, \omega_i\bar{\sigma}_{s+t}), \qquad 1 \leq i \leq n,$$

are linearly independent over $\mathbf{R}$ (Lemma 25.1.7), and therefore the n n-tuples $\omega_1\sigma, \ldots, \omega_n\sigma$ are also linearly independent over $\mathbf{R}$. Since the components of the latter are real, it follows that they form a basis for Euclidean space $\mathbf{R}^n$. Next we take in $\mathbf{R}^n$ a bi-orthogonal basis $\{e_1, \ldots, e_n\}$; i.e. a basis satisfying the condition

$$(\omega_i\sigma, e_j) = \delta_{ij},$$

where the round brackets denote the inner product in $\mathbf{R}^n$.

Let $x \in k_0$, and express x in terms of the basis $\{\omega_i\}$:

$$x = \sum x_i\omega_i, \qquad x_i \in \mathbf{Z}.$$

Obviously, $x\sigma = \sum x_i(\omega_i\sigma)$, so that by the Cauchy–Bunjakovskiĭ inequality

$$|x_j| = |(x\sigma, e_j)| \leq \|x\sigma\| \cdot \|e_j\|.$$

Hence given any ball in $\mathbf{R}^n$, there are only finitely many possible values for the integers x_j, such that $x\sigma$ lies in that ball, whence the first assertion of the lemma.

If now $x \in k_0^*$ is such that $x\tau$ lies in the ball $\|z\| < r$ in $\mathbf{R}^{s+t}$, then $x\sigma$ is in the ball of radius $e^r\sqrt{n}$ in $\mathbf{R}^n$, and the second assertion of the lemma follows from the first.

25.1.11. Exercise. The kernel of the restriction of the map τ to k_0^* is the set of all roots of unity contained in the field k. It is a finite cyclic group. (Hint. If $x\tau = 0$, then $\|x\sigma\| \leq \sqrt{n}$.)

25.1.12. Theorem. *The multiplicative group k_0^* of the ring k_0 is finitely generated.*

PROOF. By Exercise 25.1.11 it suffices to show that the additive group $A = (k_0^*)\tau$ is finitely generated; this we shall now do. Let $\{a_1, \ldots, a_m\}$ be a subset of A maximal with respect to being linearly independent over $\mathbf{R}$, and write

$$A_1 = \{\sum \alpha_i a_i \mid \alpha_i \in \mathbf{Z}\}, \qquad A_2 = \{\sum \alpha_i a_i \mid 0 \leq \alpha_i < 1\}.$$

Clearly each element $x \in A$ can be written in the form

$$x = x_1 + x_2, \qquad x_1 \in A_1, \qquad x_2 \in A_2.$$

Since $x \in A$ and $x_1 \in A$, we have that $x_2 \in A$. Now the set A_2 is bounded in $\mathbf{R}^{s+t}$, so that by Lemma 25.1.10 the intersection $A \cap A_2$ is finite. Hence there are only finitely many choices for x_2 (for all x), whence the index q say, of the subgroup A_1 in A is finite. Then $qA \subseteq A_1$, whence if follows that A is contained in the additive group generated by $a_1/q, \ldots, a_m/q$. Hence by Exercise 8.1.7, the group A is finitely generated, and the proof is complete.

Note that this proof yields the additional piece of information that the rank of the free abelian group A is at most $s + t$. It can be shown by subtler arguments that it is exactly $s + t - 1$, so that k_0^* is the direct product of a finite cyclic group and $s + t - 1$ infinite cyclic groups (Dirichlet's theorem).

Bibliography

1. Adjan, S. I. 1979. The Burnside problem and identities in groups. Springer-Verlag, Berlin–Heidelberg–New York. (Translated from the Russian by John Lennox and James Wiegold.)
*2. Belonogov, V. A., Fomin, A. N. 1976. Matrix representations in the theory of finite groups. Nauka, Moscow.
3. Botto Mura, R., Rhemtulla, A. 1977. Orderable groups. Lecture notes in pure and applied mathematics, Vol. 27. Marcel Dekker, New York–Basel.
*4. Busarkin, V. M., Gorčakov, Ju. M. 1968. Finite groups admitting partitions. Nauka, Moscow.
*5. Čarin, V. S. 1966. Topological groups. In: Scientific summaries. Algebra. 1964. pp. 123–160. Akad. Nauk SSSR Inst. Naučn. Informacii, Moscow.
6. Carter, R. W. 1965. Simple groups and simple Lie algebras. J. London Math. Soc. **40**, 193–240.
*7. Černikov, S. N. 1959. Finiteness conditions in the general theory of groups. Uspehi Mat. Nauk **14**, 45–96.
8. Černikov, S. N., Kuroš, A. G. 1953. Soluble and nilpotent groups. American Math. Soc. Translation No. 80. (Translation of: Uspehi Mat. Nauk **2** (1947), 18–59).
9. Coxeter, H. S. M., Moser, W. O. J. 1972. Generators and relations for discrete groups. Ergebnisse der Mathematik und ihrer Grenzgebiete, Bd. 14. Springer-Verlag, Berlin–Heidelberg–New York.
10. Čunihin, S. A. 1969. Subgroups of finite groups. Wolters-Noordhoff, Groningen. (Translated from the Russian by Elizabeth Rowlinson.)
*11. Čunihin, S. A., Šemetkov, L. A. 1971. Finite groups. In: Scientific summaries. Algebra. Topology. Geometry. 1969. pp. 7–70. Akad. Nauk SSSR Vsesojuz. Inst. Naučn. i Tehn. Informacii, Moscow.
12. Curtis, C. W., Reiner, I. 1962. Representation theory of finite groups and associative algebras. Pure and applied mathematics, Vol. XI. Interscience, New York-London.

13. Dieudonné, J. 1963. La géometrie des groupes classiques, Seconde édition. Ergebnisse der Mathematik und ihrer Grenzgebiete, Bd. 5. Springer-Verlag, Berlin–Heidelberg–New York.

14. Dixon, J. D. 1973. Problems in group theory. Dover, New York.

15. Fuchs, L. 1970. Infinite abelian groups. Vol. 1. Pure and applied mathematics, Vol. 36. Academic Press, New York–London.

16. Gruenberg, K. W. 1970. Cohomological topics in group theory. Lecture notes in mathematics, No. 143. Springer-Verlag, Berlin–Heidelberg–New York.

17. Hall, Jr., M. 1959. The theory of groups. Macmillan, New York.

18. Hall, Philip. 1969. The Edmonton notes on nilpotent groups. Queen Mary College mathematics notes. Mathematics Department, Queen Mary College, London.

19. Huppert, B. 1967. Endliche Gruppen I. Die Grundlehren der mathematischen Wissenschaften, Bd. 134. Springer-Verlag, Berlin–Heidelberg–New York.

20. Kargapolov, M. I. 1974. Some questions in the theory of soluble groups. In: Proceedings of the second international conference on the theory of groups, M. F. Newman (ed.), pp. 389–394. Lecture notes in mathematics No. 372. Springer-Verlag, Berlin–Heidelberg–New York.

*21. Kargapolov, M. I., Merzljakov, Ju. I. 1968. Infinite groups. In: Scientific summaries. Algebra. Topology. Geometry. 1966. pp. 57–90. Akad. Nauk SSSR Inst. Naučn. Informacii, Moscow.

22. Kegel, O. H., Wehrfritz, B. A. F. 1973. Locally finite groups. North-Holland Mathematical Library. North-Holland, Amsterdam–London; American Elsevier, New York.

23. Kokorin, A. I., Kopytov, V. M. 1974. Fully ordered groups. John Wiley and Sons, New York. (Translated from the Russian by D. Louvish.)

*24. Kostrikin, A. I. 1966. Finite groups. In: Scientific summaries. Algebra. 1964. pp. 7–46. Akad. Nauk SSSR Inst. Naučn. Informacii, Moscow.

*25. Kourovka Notebook. Unsolved problems in the theory of groups, Fifth edition. 1976. Akad. Nauk SSSR, Novosibirsk.

26. Kuroš, A. G. 1960. The theory of groups, Vols. I, II. Chelsea, New York. (Translation of second Russian edition by K. A. Hirsch.)

27. Magnus, W., Karrass, A., Solitar, D. 1966. Combinatorial group theory: Presentations of groups in terms of generators and relations. Pure and Applied Mathematics, Vol. 13. Interscience, New York.

*28. Mazurov, V. D. 1976. Finite groups. In: Scientific and technical summaries. Algebra. Topology. Geometry. Akad. Nauk SSSR, Moscow.

*29. Merzljakov, Ju. I. 1971. Linear groups. In: Scientific summaries. Algebra. Topology. Geometry. 1970. pp. 75–110. Akad. Nauk SSSR Inst. Naučn. Informacii, Moscow.

*30. Merzljakov, Ju. I. 1980. Rational groups. Nauka, Moscow.

31. Neumann, Hanna. 1967. Varieties of groups. Ergebnisse der Mathematik und ihrer Grenzgebiete, Bd. 37. Springer-Verlag, Berlin–Heidelberg–New York.

32. O'Meara, O. T. 1974. Lectures on linear groups. Published for the Conference Board of the Mathematical Sciences by the Amer. Math. Soc., Providence.

33. Plotkin, B. I. 1972. Groups of automorphisms of algebraic systems. Wolters-Noordhoff, Groningen. (Translated from the Russian by K. A. Hirsch.)

34. Pontrjagin, L. S. 1966. Topological groups. Gordon and Breach, New York–London–Paris. (Translation of the second Russian edition by Arlen Brown.)

35. Robinson, D. J. S. 1976. A new treatment of soluble groups with finiteness conditions on their abelian subgroups. Bull. London Math. Soc. **8**, 113–129.
*36. Šemetkov, L. A. (ed.). 1975. Finite groups. Proceedings of the Gomel seminar, 1973/1974. Nauka i Tehnika, Minsk.
37. Šmidt, O. Ju. 1966. Abstract theory of groups. W. H. Freeman, San Francisco–London. (Translated from the Russian by Fred Holling and J. B. Roberts.)
38. Suprunenko, D. A. 1976. Matrix groups. Translations of math. monographs, Vol. 45. Amer. Math. Soc., Providence. (Translation edited by K. A. Hirsch.)
39. Suzuki, M. 1956. Structure of a group and the structure of its lattice of subgroups. Ergebnisse der Mathematik und ihrer Grenzgebiete, Neue Folge, Heft 10. Springer-Verlag, Berlin–Göttingen–Heidelberg.
40. van der Waerden, B. L. 1970. Modern algebra. Vols. 1, 2. Frederick Ungar, New York. (Translated from the German by Fred Blum.)
41. Wehrfritz, B. A. F. 1973. Infinite linear groups. Ergebnisse der Mathematik und ihrer Grenzgebiete, Bd. 76. Springer-Verlag, New York–Heidelberg–Berlin.
42. Wehrfritz, B. A. F. 1973. Three lectures on polycyclic groups. Paper No. 3, Queen Mary College mathematics notes. Queen Mary College, London.
43. Wielandt, H. 1964. Finite permutation groups. Academic Press, New York–London. (Translated from the German by R. Bercov.)

* Items in the Russian language are indicated by an asterisk.

Subject Index

Index of Notations
of Classical Objects

Graduate Texts in Mathematics

Soft and hard cover editions are available for each volume up to vol. 14, hard cover only from Vol. 15